ALLE · ZEIT · WACH · 1842

Henk Koop

Forest Dynamics

SILVI-STAR: A Comprehensive Monitoring System

With 95 Figures

Springer-Verlag Berlin Heidelberg New York
London Paris Tokyo Hong Kong

Dr. Henk Koop
Research Institute for Nature Management
PB 46
NL-3956 ZR Leersum

ISBN-13: 978-3-642-75014-4 e-ISBN-13: 978-3-642-75012-0
DOI: 10.1007/978-3-642-75012-0

Softcover reprint of the hardcover 1st edition 1989

2131/3145-543210

to Professor Dr. Dik Thalen

Preface

Forestry, more than any other form of land use, basically runs on natural processes. Moreover, in many parts of the world forests play a determining role in soil conservation and watershed management. The preservation of forests and of a lasting production capacity demands that natural processes should be explicitly considered in forest management systems (Leibundgut 1959, 1978), so they cannot without concern be pushed into the background by economic or technical progress. On a world-wide scale there is a need for expertise on how to incorporate spontaneous processes in forest management systems. There is very little knowledge of long-term spontaneous forest development and dynamics. Only few data sets on time series from permanent quadrats, repeated forest stand measurements or forest profile analyses exist. Often the information available is restricted to a limited area and site type, to a limited form of management or to a limited set of forest development aspects. Forest managers usually have to rely on their personal, often empirical expertise. For all these reasons, natural forest reserves have been established as open-air laboratories for forest dynamics research in many European countries. The use of computers for data management and processing offers new possibilities for information systems that cover a wide range of forest development processes for many sites and forest types.

About 9 years ago, at the Research Institute for Nature Management (RIN), I set out on the task of developing a comprehensive forest ecological information system for management purposes. Ultimately, this system should enable the prediction of forest dynamics. The level of prediction should link up with the scale of forest management, such as the thinning of trees or groups of trees and its influence on species composition and growth of remaining trees, shrubs, herbs and mosses. The information collected should, among other things, also be suitable for fauna habitat description. Starting with complex physiological modules of stand models would not result in the detail required for individual trees and species other than trees (Dale et al. 1985). In fact, homogeneous stands are the exception rather than the rule in natural forest development, neither would gap models like the JABOWA family (Shugart 1984) offer the degree of detail required.

Therefore I did not choose the convenient way of starting with a simple model that can later be elaborated in more detail, but instead I began developing a digital descriptive model of the complex reality and building up time series of forest development at different sites. To come to a predictive model I carefully chose a system hierarchy for description (sensu O'Neil et al. 1986), attuned to the scale of management and the practical questions that had to be answered. The detail and precision of description has been worked out in an iterative process of feedback between description and the level of explanation required for management purposes. As steps toward the predictive model, on the basis of the data collected, application programs were developed to solve specific problems. These application programs allow causal explanation of pattern and processes at one hierarchical scale by pattern and processes at adjacent system levels. Some specific forest problems had to be tackled first.

Firstly, a forest cannot easily be observed from above and at the same time be examined from the inside, as can be done, for instance, in a heathland vegetation by simply bending down. To create a link between detailed single-tree groundtruth and aerial photographs, a model was worked out to simulate an aerial view with data collected on the ground.

Secondly, the temporal aspects of tree growth and forest-pattern development go beyond the very imagination of a human spectator. Therefore I developed a program that uses the information stored in tree rings to reconstruct and visualize forest development.

Thirdly, light is one of the most important competition factors in forests, where life-forms are adapted to striving to reach the light. So a model for light throughfall has been developed to find causal relations between the occurrence of species, their available resources (radiation) and their growth. This program can be seen as a first step toward predicting further forest development.

Acknowledgements. The Research Institute for Nature Management (RIN) supported and gave direct assistance in the preparation of this book. I owe most to Professor Dr. Dik Thalen for his special inspiration and encouragement and his numerous constructive comments on the draft of this book. Much too soon, he died suddenly on 25 December 1988 at the age of 45. As the head of the department of botany of the RIN, he was the one who took the initiative for a forest monitoring project and supported this study and its preparation as a thesis for my graduation at the Agricultural University in Wageningen (UW). He would have been one of the two promotors at my graduation at the UW.

I am thankful to Professor Dr. Ir. R.A.A. Oldeman of the department of Forest Ecology and Silviculture of the UW for his never-ceasing and stimulating support as my promotor at this graduation. Already during my studies, he created room for my special interests to develop.

Thanks are due to Ir. Ronald van de Winckel, Ir. Rienk-Jan Bijlsma, Drs. Arno van der Kruis, Drs. Henk Siebel and Drs. Johan Gudden, who served as conscientious objectors at the RIN and added to the progress of this book. Rienk-Jan Bijlsma played an important role in working out computer programs and critically following the developments of the project. Without his support the programs FOREYE and CANOPIX would not have been run.

Numerous other individuals have helped: Ir. S. Van der Werf, Drs. G. van Wirdum, Dr. G. Londo, Prof. Ir. M.S. Elzas, Prof. Ir. A. van Maaren, Prof. Ir. D.A. Stellingwerf, Dr. G.F. Peterken read preliminary drafts of the manuscript and made numerous enjoyable comments. Ir. A. Bosch assisted in solving the technical problems of wordprocessing and A. Griffioen prepared the figures in the text.

My thanks go to Ir. Henk Ruissen and the students Karl Blokland, Ryan Bevers, Jaap Wisse, Jeroen Vandeursen and Luc de Kort, who did the hard work of taking tree ring samples and measuring them. Other students carried out part of the field work: Hans van de Bos, Beatrix van Baren and Peter Hilgen.

I am especially grateful to Christel and Annemarie, who took for granted the time I spent preparing this book and disregarding normal family life.

July 1989 Henk Koop

Contents

Part A Design and Structure of the Monitoring System

Part B Data Processing

Part C Examples of System Application

Part D The Potential of SILVI-STAR

Part A Design and Structure of the Monitoring System

Chapter 1

Introduction

1.1 Problem Analysis

Forests are the end stages of a development of natural vegetation and they maintain themselves even in the face of human interference (Ellenberg 1978). Forestry, more than any other form of land use, basically and all but exclusively, rests on natural processes. The fundamental structural and operational relationships recognizable in natural forests go back to the same principles that apply to forestry plantations (Brünig 1981). Therefore, in any type of forest utilization, enough attention should be paid to incorporating of natural processes in forest management practice.

1.1.1 Over-Use of Natural Forests

Until now natural forests have survived in those parts of the world, that are far away from densely populated areas. However, during the past decades, the remaining area of all natural forests, not only the tropical rainforest, has declined enormously. Forests play an important role in conservation of species richness, but at the same time they must provide growing local populations with fuel and meet the increasing needs of the market for timber and pulpwood for the industrialized world. Natural processes are often ignored for short-term economic profit. Natural forests are exploited for valuable trees without consideration of the sustainability of resources, no matter whether it is done by exploitation companies, in irresponsible pursuit of direct gains, or by local people in the daily struggle for life. In many places this led and leads to total forest destruction. Because forests play an essential role in soil conservation and watershed management and probably even, on a world scale, in the stabilization of the atmospheric balance, forest destruction in many cases has an impact on land use far beyond the forest boundaries. Besides political action, alternative management systems for the sustainable use of the complex forest ecosystems (e.g. de Graaf 1986) must be elaborated to stop this disastrous development. Therefore knowledge of these natural processes is indispensable.

1.1.2 Ignorance of Natural Processes in Man-Made Forests

Reafforestation results in man-made forests that are often managed at a high cost of regeneration, much like an agricultural crop. As man-made forests mature and spontaneous developments can be put to use, less intensive human intervention is necessary for their maintenance.

An example of a continent where progressive forest destruction and recent reafforestation has led to a highly artificial forest cover is Western Europe, with the extreme of a purely man-made forest area in The Netherlands. During mediaeval settlement, here the wooded alluvial zones along the rivers that traversed an extensive area of treeless mires in the West and the North were cleared. Since the Bronze Age, forest devastation by grazing on higher sandy soils has led to an open landscape of heathlands with few isolated woodlands. Continued overgrazing favoured wind erosion, and the area covered by inland wind-blown sand dunes devoid of any vegetation expanded. The few remaining woodland areas were protected, either as hunting reserves or as communal property. These old woodlands were mainly managed as coppice or coppice with standards and in few cases as high forest (Buis 1985). Much of it has now been transformed into forest plantations. In the late eighteenth and nineteenth century, the heathlands and inland dunes were reafforested by private initiative and from 1899 until 1940 by the National Forest Service. As can be clearly seen from the results of the Fourth National Forest Inventory (Dirkse 1987; Dirkse and Thalen 1987), most of the present 9 % forested area has a man-made physiognomy, typical of first and second generation plantations. The second and following generation of forest plantations in The Netherlands is now maturing. Gradually, forest species are recolonizing the forests and trees are starting to regenerate spontaneously. In the past decade there has also been increasing social demand for more natural forests to favour wildlife habitats but also for ecological stabilization. This demand was stressed after the disastrous effects of two storms in 1972 and 1973 that blew down some thousands of hectares of man-made forests. As one of the fundamental principles of future forestry in The Netherlands, the long-term National Forest Plan (Anonymous 1985) states that management measures should follow natural processes and structures as much as possible. However, forest management is not yet prepared to take full advantage of natural processes.

Questions about interactions between trees have to be answered, including questions of thinning and regeneration cuttings, e.g. to what extent is spontaneous thinning in line with management goals ? How do individual trees mutually affect each other ? What mixture of tree species can coexist on a certain site type ? In what phase of stand development does competition between trees have its decisive impact for the ultimate tree species mixture ? What is the impact of spatial positions of trees and especially tree crowns upon individual tree growth and tree species coexistence ?

Other questions are related to the tree-undergrowth interaction and can be exemplified as: What is the effect of a local forest structure upon lower trees or shrubs, forest floor vegetation, and the germination of trees ? What is the impact of the size of canopy gaps upon the spontaneous regeneration of trees on different site types ? What is the interaction between the development of the herbaceous vegetation and the spontaneous regeneration of trees in these cases ? In both man-made forests and semi-natural or natural forests, preservation of the forest and of a lasting production capacity requires that natural processes be explicitly considered in forest management systems so that they cannot arbitrarily be pushed into the background for economic or technical pretexts. Moreover, good use of what nature provides freely can save money. The ancient Central European forestry tradition is a unique example of integration in management systems of regeneration processes as they take place in natural forests. The scale in space and time of the regeneration regimes mimic the patterns and sequences of phase transitions of natural forests. At the same time, this forestry tradition has outlived many movements in the market, probably because of its flexibility in adapting to changing demands (Leibundgut 1978a; Mayer 1984).

1.1.3 Restoration of Man-Made Forests as Natural Areas

When planning natural areas, e.g. in natural parks, it it must be asked, what may be expected from spontaneous forest development of woodlands that have passed through long periods of intensive utilization ? Is it necessary to take action in transformation of a current artificial forest structure when forests that have previously been managed for silviculture are planned to be ultimately left unexploited ? In what initial state should a forest structure be left unexploited to achieve the best results ? For planning in densely populated countries where claims on the land are many, it is important to know the minimum area necessary for the development of spontaneous semi-natural forests on different site types. Culturally imposed forest use, sometimes practised over the ages, e.g. coppice, coppice with standards, forest grazing or high forest, has determined the species composition of these sites (Hesmer and Schröder 1963; Pott 1981). In those areas where natural forest has vanished, the managed forest has become the last refugium for forest species. When the establishment of natural areas is considered, it is important to know what are the niches and the regeneration cycles of individual species of trees, shrubs, herbs and mosses or animals in spontaneously developing forests at different site types. Valued species for nature conservation that are found in managed forests may not find their specific niches in a spontaneous forest development cycle, and their survival may become endangered. Therefore it is necessary to know for conservation of which species nature management has to intervene in spontaneous vegetation development.

Not only in The Netherlands, but on a worldwide scale, there is need for expertise on how to incorporate natural processes in forest management systems. Knowledge of the natural processes serves not only silviculture but also management of nature preservation.

1.1.4 Lack of Coherent Information on Natural Processes

To understand the dynamic character of forest development and the role of natural processes, long-term observations are necessary. However, there is very little knowledge of long-term natural forest development and dynamics. Only few data sets on time series of permanent quadrat survey, repeated forest stand measurements or forest profile analyses exist (Bodegom 1981; Korpel 1974). Often this information is restricted to a limited area and site type or to a limited set of forest development aspects. Study of dynamic aspects of species composition meets with the problem that vegetation changes take such a long time that they go beyond the limits of a project or even beyond a human life-span (Knapp 1974). To cope with this problem, spatial vegetation differences are often used for the interpretation of vegetation changes, and/or short time observations are chained to achieve "time series" or predictions for vegetation development. In England, Rackham (1980) and in Belgium, Hermy and Stieperaere (1981) and Hermy (1985) found clear differences between species composition of ancient and more recent woodlands and could thus evaluate vegetation development in time, always supposing that at least an overall comparison of different sites and stands could be used to establish a time series. In forests, the memory represented by growth rings of trees can often help to trace back the date a regeneration unit started, and thus regeneration units of different age can be chained (Aichinger 1967; Koop 1981). The long duration of forest dynamics makes time series in forests relatively rare. They have been most successfully studied in the most dynamic phases of forest development or after a definite change in forest management, e.g. after clear-cutting (Dierschke 1978), at the beginning or the end of forest grazing (Malmer et al. 1978), or under controlled anthropogenic pressure, e.g. of trampling (Piotrowska 1978). Kauppi et al. (1978) worked out a time model for development of ground cover communities after clear-cutting. Often, ancient plots, marked in the field or on large-scale maps, were sampled again after some period by other authors. Wiegers (1985) used plots of van Zinderen-Bakker (1942) and Smittenberg (unpublished) to study succession in Dutch fen woodlands. Similar studies were reported by Christensen (1977) and Davidson and Forman (1982) over a period of 30 years, by Brewer (1980) over a 50-year period. Malmer et al. (1978) and Persson (1980) had a long time series at their disposal, records of 1916, repeated in 1935 and 1969, to reconstruct vegetation changes in a southern Swedish deciduous woodland after forest grazing had been ended. Rozendaal and Slootweg (1982), van Loon et al. (1985) and Geerdes (1985) studied 25-year old plots of Diemont (unpublished) and Van der Werf (unpublished) in The Netherlands. Bücking (1986) reports on results of permanent plots, first established by Koch and Gaisberg (1938), in a forest reserve in Baden-Würtemberg (F.R.G.) repeated every 10 years. Schmidt (1988) gives examples of yearly fluctuations of cover values, number of shoots and flowers of a herb-layer community of terminal stage beech forest on calcareous soil near Göttingen (F.R.G.). Reports on regular records of permanent plots over periods longer than 10 years are hardly available. In most cases where such plots exist, no results have been published. Most methods used for forest monitoring refer to isolated forest components. For instance, only trees, floristic composition or fauna have been studied. Most of the time data for explanatory relations between the forest components are not available. Lack of continuity in research programs, bad documentation and storage of data and lack of comparative methods have caused this lack of information on natural processes in forests (Peterken and Backmeroff 1988). To study spontaneous processes in their pure form without human interference and to ensure some degree of continuity and integration of forest research, in many European countries including The Netherlands, natural forest reserve programs have been started. In the reserves any form of internal human interference has stopped. European forest reserves were first proposed by Hesmer (1934). In the middle of the nineteenth century, already for aesthethic reasons, forests had been reserved, such as in Fontainebleau (France) and in The Neuenburger and Hasbrucher Urwald (F.R.G.). Forest reserve programs in the F.R.G. have been described by Lamprecht (1969), Trautmann (1969), Wilmanns (1977), Genssler (1979) and Bücking (1986). A somewhat similar program is conducted in the D.R.G.: Scamoni (1953), (Niemann 1968). In some other Central European countries also forest reserve programs have been started, e.g. in Switzerland (Leibundgut 1966), Czechoslovakia (Vyskot et al. 1981; Korpel 1982; Prusa 1985) and Austria (Zukrigl 1983). Five British semi-natural forests were recorded for more than 25 years (Peterken and Jones 1987; Peterken and Backmeroff 1988). Other integrated forest research projects were carried out as part of the International Biological Program (e.g. Newbould 1967; Duvigneaud 1971; Bouchon et al. 1973; Falinski 1977, 1986).

1.1.5 Tools for Prediction

Expertise with regard to the incorporation of natural processes in forest management requires the ability to predict what development can be expected in forests in different states and at different sites. Different procedures can be followed to make predictions. Mostly predictions concerning forest dynamics are based on best personal judgement of managers or scientists. Reliability of that personal judgement is enhanced by long-standing experience, by the communication of experience in circles of experts and by the human memory. In our dynamic modern world, where time is short, very few people have the opportunity to observe forests for longer time periods. Forest managers and decision makers spend most of their time behind their desk or they move after a few years to another office, without having been able to see the effect of their decisions. Times when personal experience could be optimally transmitted to successors, as the job of a forester passed from father to son, seem to have vanished for ever, in any case as concerns the transfer of knowledge. The use of databases as a basis for prediction can help to overcome the limitations of individual human memory and human communication. However, not all experience can be transmitted in a digital form. Large tables of numbers cannot replace or be translated into the vivid pictures of forest states stored in human memory. The risk of building up extensive databases is that they end up as "data cemeteries". The data analysis of time series, or a reconstruction of forest development, allows the assessment of the probability of developments. This assessment, however, includes a final step of personal judgement.

When trying to develop predictive models, it is important to know the nature of the forest vegetation to be modelled. Prior conditions have to be given in function of what should be predicted. Dale et al. (1985) state that "No model will ever predict all of the forest attributes precisely due to complexity and variability of ecological factors affecting tree growth. The best models will explain the model variation from the data and be appropriate for the questions at hand whether they relate to forest management or ecological succession". The heterogeneous nature of the vegetation of natural forests has been described by many authors, e.g. Anonymous (1958), Zukrigl et al. (1963), Barkman (1968, 1970, 1979), Glavac (1972), Zukrigl (1973, 1978), Oldeman (1974a, 1978, 1983, 1989), Koop (1981, 1982) and Prusa (1985). They pay attention to the problems that arise when using traditional methods of phyto-sociology, especially for permanent plot studies. The authors focus on the patchy character and dynamics of forests in the tree compartment, causing heterogeneity in species composition of forest communities. The structure of the higher layers determines the patchy character of the lower layers (Ellenberg 1939; Eber 1972). At a more detailed scale level, heterogeneity may arise, e.g. when uprooted trees leave a mouldering dead trunk and a mound-and-pit micro-relief (Londo 1977; Beatty 1981; Koop 1981). Not only spatial heterogeneity was identified as a cause of problems, but also the occurrence of sequential or successive phases. Watt (1947) recognized genetically closely related phases as integral parts of a coordinated whole.

Ever since studies at phyto-coenosis level started, scientists have struggled with the heterogeneity of plant communities (Mueller-Dombois and Ellenberg 1974; Jahn 1982; van Wirdum 1987). Samples should preferably be homogeneous as, after classification, the derived associations also (Braun-Blanquet 1964). Godron (1971) tried to solve the homogeneity problem in stating that a vegetation is always more or less heterogeneous. He states that the statistical null hypothesis should not be homogeneity but heterogeneity. The objective of his study is to quantify heterogeneity, considering it no longer as an enemy but as an ally.

The use of computers for data management and processing offers new possibilities for the construction of predictive models that cover a wide range of forest development processes for many sites and forest types. However, there is still quite a gap between scientists who have studied the complex reality of forest architecture in a descriptive way and those who build models starting at the other end on a physiological basis with simple assumptions about forest architecture. The first group spends a great deal of time in the field, the other in the laboratory and in front of a computer screen. Most existing predictive forest models relate to the growth of no other species than trees, whether in stands or in patches that are assumed to be homogeneous (Shugart 1984; Dale et al. 1985). Relations between trees and other lower forest compartments most often are lacking. Complete simulation of complex forest architecture of natural forest is still beyond the reach of any forest model at present.

1.2 Objectives

The first objective of this study is to develop a method for digital description of forest architecture and species composition for monitoring purposes. The method must allow for the detection of changes in forest vegetation structure and species composition at different scales in space and time. Graphical data, such as maps and profile drawings, therefore need numerical expression. Standardized data collection is required. The collected data should provide a detailed, yet comprehensive, database enabling descriptive and explanatory modelling at different scales for forestry, vegetation science and fauna habitat studies. Spatial arrangement of individual trees and its effects upon gap regeneration and herbaceous vegetation development should be the first main subject for monitoring.

The second objective is to develop an information system that allows searches on many aspects, simultaneously favouring its use for ad-hoc consultancy purposes. Final judgement and application of the information system output should be facilitated. The output on forest dynamics, i.e. patterns and processes, ought to be suitable for direct visual interpretation to enable demonstration to colleagues and managers in the field. To safeguard continuity in data storage and query on a long-term basis, it is necessary to use universal and commercially available database management- and geographical information systems for storage.

The third objective is to develop computer tools for a predictive forest development model. The model should focus on the prediction of the influence of single-tree and tree group expansions and/or removals on the establishment and growth of other organisms, such as trees, shrubs, herbs and mosses in and below such trees or tree groups. The first step towards a predictive model will be a descriptive model based on the information system data, the second the development of explanatory application programs as an extension of the descriptive model.

1.3 Outline of the Text

The text of this book is divided into four parts. Part A, Development and Structure of the Monitoring System (Chapters 1 to 3), deals with the problem analysis and the objectives of the study. The development of methods for description, data storage in an information system and of models for prediction are described. Relevant literature is briefly reviewed in introductory paragraphs. This literature survey has been limited to a discussion of the techniques included in the information system; it is not intended to give an exhaustive review of the literature on forest succession.

Part B, Data Processing (Chapters 4 to 6), deals with a description of the basic tree model (Chapter 4), whereas in Chapters 5 to 8, different tools for modelling using the information system data are described.

Part C, Examples of System Application (Chapters 7 to 11), demonstrates the monitoring system at work in some forest objects under study, using the tools described in Part B. Chapters 7 and 8 deal with reconstruction of long-term or cyclic and non-cyclic forest dynamics by means of tree ring analysis, traces of uprooting and interpretation of the present forest structure. Preliminary results of forest monitoring in some selected objects under study are discussed in Chapter 9. In Chapter 10 a number of variables is studied across the objects under study. Chapter eleven deals with the results of simulations of light throughfall and their use to explain forest structure and dynamics.

Part D, The Potential of SILVI-STAR (Chapters 12 and 13), gives examples of the use of the monitoring system as a tool in forest management. In Chapter 12, the monitoring system as a tool for management practice and science is discussed, in Chapter 13, the scope of improvement and further development of the system.

Chapter 2

Design of the System

In 1980, a long-term ecological monitoring project was initiated to study development processes in forests under natural conditions (Koop 1982, 1984, 1986b, 1988). It aimed at building up an ecological information system for forest management, with special emphasis on its applicability for nature conservation purposes. Broader applicability of the system has been considered from the beginning, e.g. in the Netherlands for multi-purpose and production forestry, by incorporating spontaneous processes in the management systems. Three phases in the development of the project can be distinguished: the first concerns the working out of appropriate techniques for data collection in the field; the second the organization of data storage and retrieval; and third, data processing and interpretation. Collected data were planned to be used also for validation of predictive models of forest development.

2.1 Phase 1: Working Out Appropriate Techniques for Data Collection

Against the background of the problem analyses and objectives of the study, the spatial arrangement of individual trees and its effect upon gap regeneration and herbaceous layer development should be the main subject for monitoring and modelling. On a higher scale level, patterns in tree species' distribution and developmental phase should be studied.

Partially on the basis of earlier experience (Koop 1981), a system was worked out from already existing techniques for the analysis of structure and species composition of natural and semi-natural forests: (1) remote sensing, (2) forest profile analysis, (3) methods of forest inventory used in natural forest reserves in Central Europe, (4) permanent quadrat survey in combination with vegetation mapping and (5) ground photography.

2.1.1 Reflections on Plot Size and Frequency of Measurements

Processes in the forest take place at different scale levels in space and time. Changes in time vary from seasonal fluctuations to changes that can be observed and measured only after years, decades or even longer periods. The frequency of observation should be adapted to allow the study of these processes. Events such as the death of trees, windthrows and tree breaks are measurable over short periods. Soil disturbance by uprooting and changes in the herbaceous vegetation closely related to such abrupt changes in the forest canopy should be measured at the same intervals. For continuous, gradual processes, such as tree growth, the methods of measurement must be attuned to the precision needed to solve the stated problem. The desired interval between measurements can be determined by comparing an estimate of the rate of change and the precision that can be reached by a certain method for the variable to be measured.

Changes in patterns of plants of different size take place at spatially different scale levels. The information system concerned with complex forest-ecosystem behaviour at different scales in time and space to collect data for pattern analysis should focus on areas of different size at different frequencies. Techniques for data collection should therefore be organized in a hierarchical system concept.

The hierarchy theory (Simon 1962) is concerned with systems that have a certain type of hierarchically organized complexity. Such systems can be divided into discrete functional systems or subsystems, operating at different scales. A set of subsystems together can explain the functioning of a system at a higher integration level. One fundamental consequence of hierarchical structure is that non-equilibrium dynamics or spatial heterogeneity at one scale is incorporated into higher level behaviour, where it can be translated into equilibrium or constancy (O'Neil et al. 1986). For example, at a higher level, disturbance frequency can be considered as a constraint that governs an equilibrium species assemblage (Allen and Starr 1982). Urban et al. (1987) recapitulate that a hierarchical perspective would emphasize three strategic concerns in an analysis of landscape pattern:

1. detect pattern and define its spatial and temporal scale at a specific level;
2. infer which factors generate pattern, whether they be demographic processes, environmental constraints, disturbances or a combination of these;
3. relate this pattern to adjacent levels.

According to this theory, Delcourt et al. (1983), Urban et al. (1987) and Falinski (1988) arranged different disturbance regimes, forest processes, environmental constraints and vegetation patterns. They differ in their spatial and temporal extent in logarithmic scaled space-time domains (Fig.2.1). The use of hierarchically organized subsystems implies hierarchical interactions.

This allows causal explanations by explaining patterns and processes on a certain scale by patterns and processes at a lower hierarchical level. For the individual subsystems, the

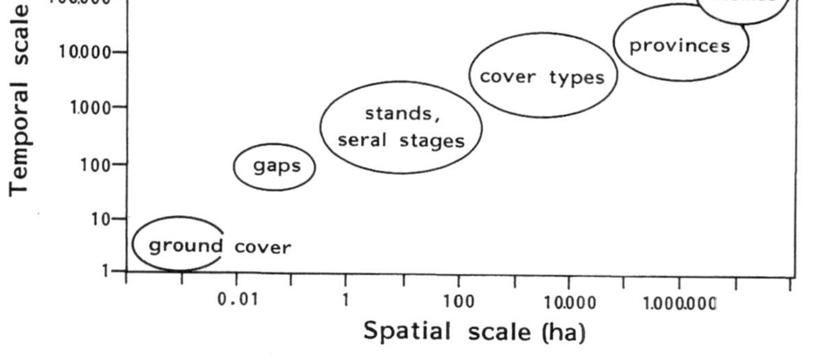

Fig. 2.1. Vegetation patterns, viewed in the context of space-time domains. (After Delcourt et al. 1983)

quantity of data will be less numerous and need not be numerical. Statistical proof of observed phenomena will be difficult. The use of fewer subsystems and higher quantities of data per level, on the other hand, improves the possibilities for statistical analyses of data; these data have to be numerical (Oldeman 1989). A system state statistically described in terms of means and standard errors (Husch 1971) at two moments quantifies a shift. The process responsible for the shift cannot be explained, because hierarchical levels in such an approach are lacking. The number of levels distinguished and their delimitations can be freely choosen, depending on the requirements of the study concerned.

2.1.2 A Hierarchy for Forest Analysis

The first visible impact of most silvicultural treatment is structural. Forest structure made up by trees can easily be depicted in maps and profile drawings and can thus be made accessible to managers in the field. Therefore the use of structure in hierarchy as the prime criterium would be very suitable for the purpose of this study. Oldeman (1974a, 1978, 1989), studying spatially complex tropical rainforests, developed a hierarchical system, adapted to the dynamic and patchy character of natural forest vegetation. He aimed at giving his hierarchical levels spatial limits. At each level biotic as well as abiotic subsystems can be distinguished. His levels will be discussed below against the objectives and requirements for the information system (Chap. 1).

The Tree Compartment
The space occupied by one single tree during its lifespan is defined by Oldeman (1974b, 1983) as the tree ecotope. Morphological criteria for the observation of trees are integrated into whole tree architecture and growth models, which can be considered as differentiated branching complexes (Hallé et al. 1978). Subsystems of the tree, as proposed by Zimmermann and Brown (1971), Edelin (1977), Hallé et al. (1978) and De Reffye (1979) as organs integrated into differentiated organ complexes, are beyond the scope of this study, considering its objectives (Chap. 1).

The Regeneration Unit
Oldeman (1983, 1989) defines the regeneration unit or eco-unit as: "every surface on which at one moment in time a vegetation development has begun, of which the architecture, ecophysiological functioning and species composition are ordained by one set of trees until the end". According to Knapp (1974), regeneration unit development can be treated as a cyclic subprocess in "succession". The regeneration unit level fits the scale of management interventions of cutting and thinning, where by manipulation the interaction between trees and the surrounding stand and the interaction with spontaneous regeneration of trees and herbaceous species are being influenced.

The Non-Woody Compartment
For two reasons there is the need to distinguish, besides a woody, a non-woody compartment of the regeneration unit.
1. Soil disturbances by uprooting of trees, decaying tree stems and sun flecks cause heterogeneity in the forest floor vegetation pattern within the regeneration unit.
2. The occurrence of mostly non-woody life forms may be highly determined by trees. The replacement of the trees thus automatically entails the death of the forest regeneration unit as a whole, and its replacement by one or more other units. But

patches of herbs or lianas in the new unit may also determine the establishment and growth of trees. Therefore in terms of hierarchy, a subsystem of non-woody species of equivalent hierarchical position as the tree is needed.

According to the concept of synusiae (Barkman 1973), groups of plant populations can be distinguished that belong to one life form or are separated from other populations in time or space or by a fundamentally deviant use of the site. This concept covers the need expressed above and can be added to Oldeman's hierarchy.

The Mosaic of Regeneration Units: the Silvatic Mosaic

A set of regeneration units, often but not always of different ages, builds a forest mosaic. Natural forests consist of a continuously shifting mosaic of regeneration units. Because the term forest mosaic was already used in other models in a deviant way, e.g. landscape models (Shugart 1984), the term sylvatic unit or sylvatic mosaic was introduced for this particular biotic mosaic (Oldeman 1983, 1989). The sylvatic mosaic is assumed to occur on one homogeneous site type. Mosaics with patterns of different coarseness have been described for various sites (Koop 1981; Korpel 1982). At this hierarchical level, minimum areas of the mosaics of regeneration units have to be estimated. The coarseness of the sylvatic mosaic gives information for transformation management.

The Site Mosaic

At the highest integration level different sylvatic mosaics on different sites can be integrated in a site mosaic. However, site patches, in a fine site mosaic, may be smaller than the sylvatic mosaic area that shows durable proportionality of regeneration unit age classes. In this case the sylvatic mosaic level can be skipped. The interactions between regeneration units of the same site can still be studied if the size and the ratio of surface to boundary length of separate site patches are large enough. To attain the surface of one complete sylvatic mosaic, a number of site patches scattered over the site mosaic have to be lumped together. In the latter case, in sequence of just size hierarchy, the sylvatic mosaic and site mosaic ought to be interchanged, although this is not so in terms of complexity.

The level of the site mosaic is beyond the scope of the study unless this level is interchanged with that of the sylvatic mosaic, as mentioned above. The separate units of the site mosaic, i.e. sites or sylvatic mosaics, are used for the selection criterion for the study areas to be included in the system.

2.1.3 Nested Plot Design

The levels distinguished in the above-described hierarchy correspond very well to the scale levels that are most relevant for management purposes. The regeneration unit, its subsystems (the non-woody and tree compartment) and its supersystem (the sylvatic mosaic) are the hierarchic levels on which the information system has to concentrate. Therefore the scale levels of data collection have to be attuned to the surface area that these hierarchic levels cover in the field.

Minimum Area Needed for the Sylvatic Mosaic

When studied over a sufficiently large area, the sylvatic mosaic proved to show a durable proportionality of age classes of regeneration units. Koop (1981) and Korpel (1982) independently distinguished a minimum area required for the permanent dynamic equilibrium of a sylvatic mosaic. In this so-called minimum structure area (Koop 1981), all developmental phases should be permanently present, each of these phases occupying a part of the entire area proportional to the duration of the individual developmental phases from the entire developmental cycle. The durable proportionality of age classes of regeneration units was for the north German acid and mesotrophic beech forests (*Fago-Quercetum* and *Milio-Fagetum*) and the oak-hornbeam forests (*Stellario-Carpinetum*) reached on areas of 40, 25 and 10 ha, respectively (Koop 1981). In Slovakia for beech forests (*Fagetum typicum*), for mixed fir-beech forests (*Fagetum-abietosum* and *Abieto-Fagetum*) and for spruce forests in the high altitude zone (*Sorbeto-Piceetum*) according to Korpel (1982) this area amounts to 30, 40-50 and 60-70 hectares, respectively. When studying the sylvatic mosaic, the area monitored should at least equal the minimum structure area.

Minimum Area for the Regeneration Unit

The detailed architectural study of one regeneration unit and its individual trees requires some hundreds of square metres, the exact surface depending on the build-up of the sylvatic mosaic and its subsystems (Oldeman 1979). Regeneration unit size in natural forests in Europe seldom exceeds a diameter of one to two tree heights (Koop 1981; Korpel 1982; Lemeé et al. 1986). In a wide variety of transects from all over the world, Oldeman (1989) distinguishes regeneration units of about similar size. The shifting mosaic character of regeneration units and the non-woody compartment necessitates monitoring the shifting exterior borderlines and interior sampling of the units.

Minimum Area Needed for the Non-Woody Compartment

In general, vegetation sample plots should cover the minimum area of the vegetation type they represent. An area of 100 to 400 m^2 is thought to be large enough for the forest species composition sample in Europe (Mueller-Dombois and Ellenberg 1974). For the analysis of influences of sun patches, uprootings or dead logs, at the scale of the microcoenose, Glavac (1972) suggests not to use a plot size of 400 m^2, but instead many smaller plots of 1 to 4 m^2 arranged in a particular pattern. This offers the opportunity to study local developments that might be averaged out when quantifying information in one big plot. To obtain a sample that covers the minimum area, the small plot values can be combined. For research in natural forest reserves in Western Germany, for similar reasons Jahn (1980) proposed to note the presence of species on the points of intersection of a grid within a 100 m^2 vegetation plot.

Taking into account the advantage of standardized metric plot size and the natural limits of hierarchic units, an optimal sample plot size for each hierarchic level was selected. Each of the plots of standardized size should be large enough to enclose the hierarchic levels mentioned above.

In this context it is important to realize that the impact of high trees upon lower vegetation components reaches further than that of lower ones. Tree canopy gaps outside a plot can influence species composition inside the plot because of influencing oblique radiation (Koop 1981). To explain changes in a plot in which small life forms are sampled, data on trees outside the actual plot are indispensable. Bigger life forms therefore need to be sampled in bigger plot size. This is in line with the increasing size of minimum areas of vegetation types described for only mosses, herbs, understorey trees or shrubs and trees (Oldeman 1979, 1980). The principle of the nested plot design, with overlapping plots of increasing size and decreasing observation intensity was incorporated to eliminate this problem (Fig. 2.2; Table 2.1).

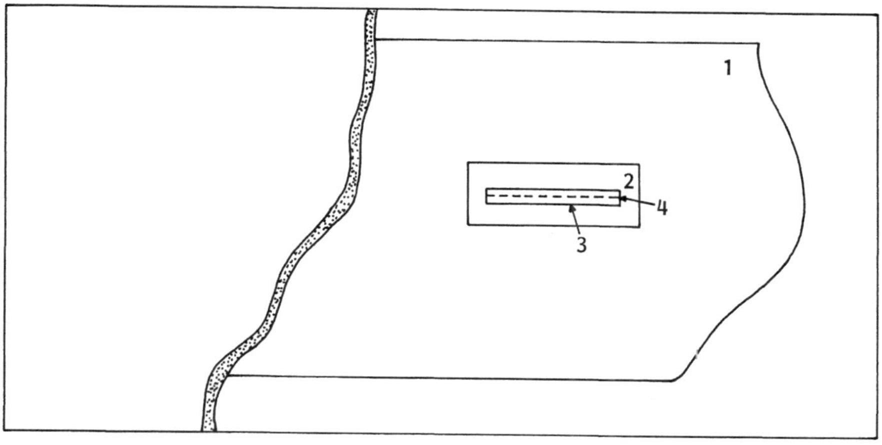

Fig. 2.2. The principle of nested plot design worked out for four scale levels. The forest type area (*1*) covering the minimum structure area (msa) (Koop 1981) is surveyed with 10-year intervals with remote sensing techniques. A rectangular core area (*2*) is inventoried at the same frequency. For low forests an area ranging only one tree height around the woody transect area (*3*) is inventoried. It is used as a groundtruth reference for the forest type area and is meant to provide a buffer zone around the woody transect area. Information on forest structure and dynamics from the core area is used for explanation of changes in floristic composition in the herbaceous transect area (*4*)

Table 2.1. Description of the scale levels, their dimension and area and the frequency for reassessment. *msa* is the size of the minimum structure area

	Description	Dimension	Area	Frequency
1.	Forest type area	-	msa	10 years
2.	Core area	$70 \times 140\ m^2$	1 ha	5-10 years
3.	Woody transect area	$10 \times 100\ m^2$	0.1 ha	5-10 years
4.	Herbaceous transect area	$2 \times 100\ m^2$	0.02 ha	3-5 years

The mosaic of units described for one hierarchical level can be charted, at a scale level large enough to hold several units, while the nested plot design enables interior sampling at the scale level below. System behaviour at one scale level can be explained with the help of information from hierarchically lower levels. The effect of trees upon the herbaceous vegetation, in the central strip transect, can only be interpreted with knowledge of forest structure in the core area, covering a surface extending at least one maximum tree height around the transect. The size of the core area was therefore made conditional to the natural measure of one maximum tree height. The already existing

Table 2.2. Data collection techniques incorporated in the information system fitted in the nested plot design for both interior (I) and exterior (E) description of the hierarchical system levels

Hierarchical system levels									
Non-woody compartment	I		E						
Woody compartment		I		E	E	E			
Regeneration unit			I	I	I	I	I	E	
Sylvatic mosaic							I	E	
Site mosaic							I	I	E
Data collection techniques / Scales of data collection	Adjacent 2 x 2 m Pq	Profile drawing	Vegetation map 1:200	Tree measurements	Tree charting	Ground photography	10 x 10 m Pq	Forest structure map 1:2500	Vegetation map 1:2500
Herbaceous transect area	*								
Woody transect area		*							
Core area			*	*	*	*	*		
Forest type area							*	*	*

techniques of data collection mentioned earlier have been fitted in the nested plot design (Table 2.2).

2.2 Phase 2: The Organization of Data Storage and Retrieval

For data storage, universal and commercially available databases have been used. For alphanumeric data ORACLE (ORACLE Corporation 1987) and for map information the ARC/INFO geographic information system (ESRI 1987) and the ERDAS-package (ERDAS 1988) for image processing has been used. Both systems are standards for use on DEC VAX systems within the institutions of the Ministry of Agriculture in the Netherlands.

At the start of the project all data were collected on field-forms or as drawings. Later, the use of a hand-held computer in combination with a tree calliper and a hypsometer enabled direct data input into files in the field. Only maps are first hand-drawn and then digitized. In this phase programs for graphical output from the database of plans and profiles of forests have been elaborated. The basic tree model was developed here.

2.3 Phase 3: Data Processing and Interpretation

Models are not universal but restricted to the specific task they are built for. Chaining physiological growth modules of individual plant species (e.g. of Douglas fir by Mohren 1987 of different-sized growth forms and architecture models sensu Hallé et al. 1978) to build a multi-species competition model seems an impossible task. The nature of the forest vegetation presents some specific problems to be solved.

Horizontal and vertical heterogeneity is characteristic for natural forest development. Forest patches can be distinguished that have a temporary nature. Differences in space may have an importance similar to differences in time. Existing forest growth models hypothesize homogeneity of a stand or at least of arbitrary subunits (Botkin et al. 1972). They attempt to predict for one homogeneous stand or for a set of patches but not for the individual patch or even the individual tree.

Nevertheless, the basic elements in a predictive model to solve problems put forward in Chapter 1 should fit the scale of management interventions: i.e. the interaction between individual trees and the surrounding stand. If horizontal heterogeneity and individual trees are part of a model, most of the time it is presumed that vertically the forest is mono-layered (e.g. Faber 1983).

However, the vertical dimension of the forest complicates elaboration and application for forests of growth models developed for monolayered homogeneous crops. For example, the light extinction in forest stands or even in homogeneous patches is different from that in a homogeneous crop because of the clumped occurrence of tree crowns. In fact, the forest is built up by separate crops arranged in space, i.e. tree crowns, not to speak of lianes that cover tree crowns and function as curtains at the edge of canopy gaps. This vertical aspect always causes interactions with neighbouring patches. The scale of disturbance is caused by the death, break or uprooting of a tree. All causes of tree extinction are influenced not only by tree age but also by the structure of the forest itself, e.g. shading and roughness of the canopy and position of a tree in a wind-exposed or sheltered environment. The mechanical aspects of destruction by falling trees are responsible for the scale and pattern of disturbance and regeneration (Oldeman 1974a). Different-sized forest components like trees, shrubs, herbs and mosses stress the need for a hierarchical organized approach.

Starting with complex physiological modules of stand models would not result in the detail required for individual trees and species other than trees. Although growth can be assigned to individual trees and tree species composition can be predicted, gap models like the JABOWA family (Shugart 1984) would also not meet the detail required.

In an iterative simulation model, the casual nature of, for example, trees that in their fall cause damage to other trees, or trees that uproot or rot in the heart and break, make predictions of individual tree growth almost impossible. The aspects of damage caused by the fall of a tree and the light interception of trees in a complex forest structure necessitate three-dimensional modelling. The complex reality of natural forest developments in details of spatial arrangement of trees of different size and shape could not as yet be simulated with existing models.

Therefore not the way of starting with a basic model, that can be elaborated in more detail later on, is chosen. Instead a descriptive model of the complex reality was developed and time series of forest development at different sites were built up. On the basis of the data collected, applicative computer programs were developed to solve specific problems on the way towards a predictive model (Chap. 6).

Because light is the most important competition factor in forests, a model for light throughfall has been developed to find causal relations between occurrence of species and their growth. Relations between light simulation and vegetation composition processing is elaborated, and correlative studies have been carried out to find the relation between tree ring growth and simulated light values. These programs can be seen as a first step towards predicting further forest development.

To enable the link between detailed single-tree ground-truth and aerial photographs, a model to simulate an aerial view with data collected on the ground has been worked out. Because tree growth and forest pattern development exceed the imagination of a human spectator, a program that uses the information stored in tree rings to reconstruct forest development has been developed. This program enables visual interpretation and can be used for demonstration for colleagues and managers in the field.

Chapter 3

Data Collection and Data Storage

In the preceding chapter the reasoning behind the decision to adopt a nested plot design with four scale levels of data collection was given. In the present chapter the data collection techniques for the different scale levels (Table 2.1) will be described. The scale levels will be dealt with, starting with the highest level of the forest type area and ending with the herbaceous transect area. How these scale levels fit in the hierarchical system level of the forest site mosaic will be dealt with. Data storage will be discussed.

The hierarchical level of the forest site mosaic reaches beyond the nested plot design considered here. Only in the case of a mosaic or complex of small site patches, does the nested plot set up focus on more than one site unit. In all other cases, the site type is the selection criterion for the forest objects to be included in the studies carried out with the information system and the key for extrapolating the results to other forests of the same site type.

Until now, for characterization and classification of the sites, the concept of the PNV (Potential Natural Vegetation) is used. Tüxen (1956) based the concept of PNV on the genesis of actual but secondary vegetation types in man-made landscapes. The author defines the PNV as the species composition that would develop, under present existing abiotic circumstances, without human interference and within a time span over which the vegetation can reach its end state. The PNV concept is a tool for recognition of an actual vegetation type as part of a developmental series. These series are named after their oldest present development stages (Kalkhoven and Van der Werf 1988). This approach helps to obtain a generalized picture from the multitude of actual species combinations, in space and in time (Ellenberg 1978).

The Potential Natural Vegetation of the Netherlands was mapped by Van der Werf, in Trautmann (1972) for the Dutch-German frontier zone, by Kalkhoven et al. (1976) and Van de Brink and Van der Werf (1977) for the Veluwe region. Nordrhein-Westfalen (FRG) was mapped by Trautmann (1972) and part of Niedersachsen by Krause and Schröder (1979) scale 1:500,000. Géhu (1973) worked on the Potential Natural Vegetation of northern France. The whole of Europe is mapped to scale 1:3000,000 by Noirfalise (1987).

Recently Van der Werf (1989) worked out the classification of potential natural vegetation types for The Netherlands in more detail. Early drafts of this classification were the basis for the selection of forest objects to be included in the monitoring system. For each PNV type of the Netherlands, the forest object closest to the end stage of the PNV was selected. Species composition should be representative for the PNV and as far as possible soil profiles should be undisturbed. Ancient woodlands were preferred unless the PNV is characteristic for younger woodlands. If selection for a PNV within The Netherlands was unsatisfactory, for reference the most natural representative of the PNV was sought abroad. In the Appendix all objects now under study are listed.

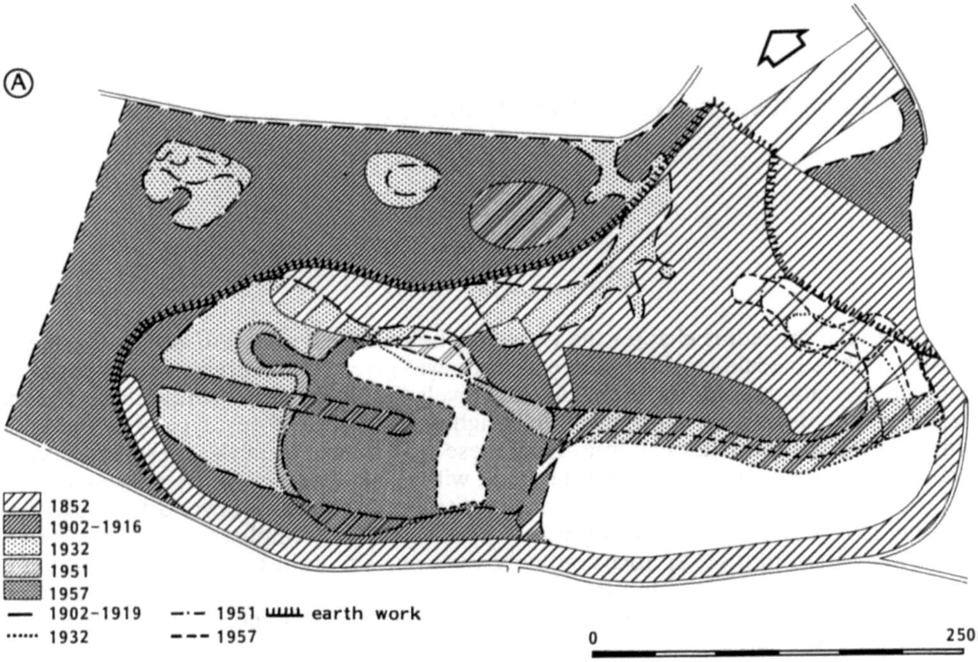
A
1852
1902-1916
1932
1951
1957
1902-1919
1932
1951
1957
earth work
0
250m

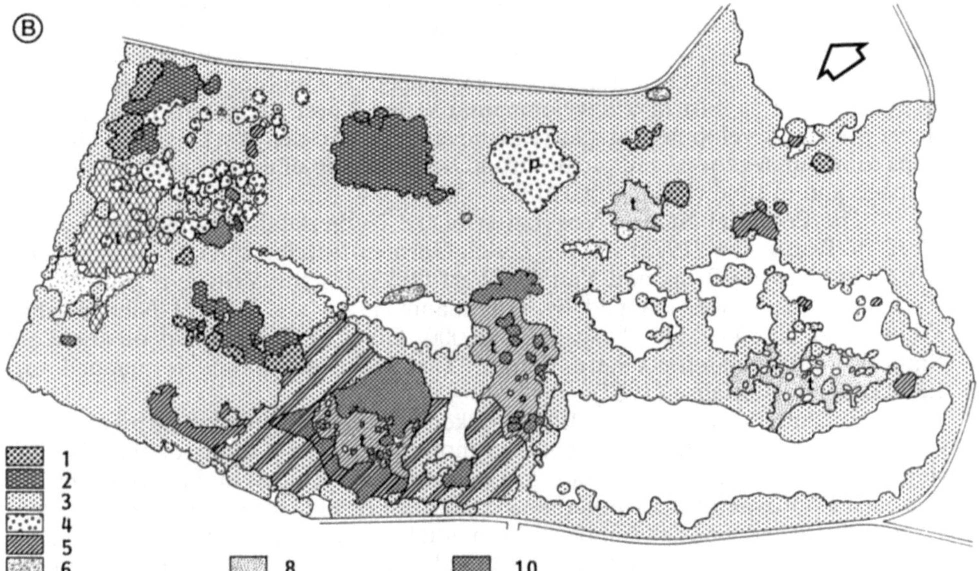
B
p
t
1
2
3
4
5
6
7
8
9
10
11
12

3.1 Forest Type Area

The level of one forest type covers one sylvatic mosaic with its characteristic mosaic of regeneration units. At the area, covering at least one "minimum structure area", the scale of regeneration of one forest type can be studied. To monitor vegetation pattern changes at the scale level of the sylvatic unit, an area of 10 to 40 ha around the core area is mapped from large-scale aerial photographs (1:4000 to 1:8000). The spatial patterns of vegetation composition and vegetation structure are recorded every 10 years. Aerial photographs in true colour, colour infrared, or both, are digitized by means of a video camera or scanner, and image files are kept on tapes. Besides digital storage, visual photo interpretations, such as vegetation maps and forest structure maps, are kept on shrink-proof tracing sheets and stored in the same way as the chartings of the core area.

3.1.1 Mapping of Forest History

Information on forest history can be gathered from old maps. In The Netherlands a regular ordnance survey 1:50,000 and later 1:25,000 has been carried out since 1850, covering the entire country (Van der Linden 1973). Older maps originating from the military engineers' archives date back to the second half of the eighteenth century. They do not cover the entire country and their reliability varies (Koeman 1963). The first land registry maps in the period 1824 to 1848 are very interesting. On these maps, all cadastral units are charted at scale 1:5000. Landownership and use are registered. In fact, these maps were used as a basis for the first ordnance survey map. A useful source of information for more recent history is aerial photography. Written documents in the form of articles, books, etc., but also information from archives and personal communication from local people may help to solve questions about former local land use. Rackham (1976, 1980) and Peterken and Jones (1987) mapped ancient woodlands by collating historical information sources with mapping of earthworks that mark former woodland boundaries.

In the developed monitoring system, a forest history map is made at scale 1:2500. Earthworks are mapped and the age of the woodlands is mapped as reconstructed from old maps and written documents (Fig. 3.1a).

Fig. 3.1.A Example of forest history mapping of the forest type area of the Weenderbos (province Groningen, The Netherlands). **B** Example of forest structure mapping of the forest type area of the Weenderbos. The bigger part of the reserve is covered by an old tree phase (height > 20 m) in which tree species are distinguished. Letters *p* and *t* indicate pole tree phases (height < 20 m) and thicket phases (height < 10 m). *1 Fagus sylvatica, 2 Quercus rubra, 3 Quercus robur, 4 Fraxinus excelsior, 5 Betula pubescens, 6 Populus tremula, 7 Acer pseudoplatanus, 8 Corylus avellana, 9 Sorbus aucuparia, 10 Alnus glutinosa, 11 Salix cinerea, 12* open treeless area

3.1.2 Mapping of Vegetation Structure

A study on forest structure in terms of patterns in stands was probably first published by Watt (1947), on beech stands in England. He stated that recognition of genetically closely related phases as separate entities may be convenient in classification, but successful practical management of such woods depends on their recognition as integral parts of a coordinated whole. For the purpose of mapping forest mosaics, many efforts have been made towards classification of regeneration units. Phase definitions have been published by many authors: Niemann (1968), Leibundgut (1959, 1978b, 1982), Houtzagers (1954), Hillgarter (1971, 1976), Bouchon et al. (1973), Knapp (1974), Whitmore (1975), Mayer (1976), Cochet (1977), Derkman and Koop (1977), Hallé et al. (1978), Oldeman (1974a, 1978, 1979, 1989), Bormann and Likens (1979), Koop (1981), Torquebiau (1985, 1986), Korpel (1982), van Baren and Hilgen (1984) and Koop and Hilgen (1986).

Table 3.1. Examples of phase descriptions. N is the number of phases distinguished. Note that not all phases refer to the same hierarchic level

Authors that refer to regeneration units

Author	Year	N	Nomenclature
Watt	1947	3	Building, Mature, Gap
Houtzagers	1954	4	Dense, Thicket, Poletree, Tree
Bouchon	1973	5	Open, Dense, Thicket, Poletree, Tree
Oldeman	1974a	3	Growing, Homeostatic, Chablis
Whitmore	1975	3	Building, Mature, Gap
Koop	1981	6	Bare-Dense, Thicket, Poletree, Tree, Decay
Korpel	1982	3	Growing up, Optimum, Destruction
Torquebiau	1986	7	Chablis, Phase of the future (1L-1H-2L-2H), Phase of the past

Authors that refer to regeneration unit complexes

Author	Year	N	Nomenclature
Mayer	1976	5	Youth, Initial, Optimum, Terminal, Regeneration
Hillgarter	1976	4	Initial, Optimum, Terminal, Regeneration
Leibundgut	1982	7	Youth, Homogeneous, Poletree and Tree Phases Optimum Ageing Decay Regeneration, Plenter

Authors that refer to the sylvatic unit

Author	Year	N	Nomenclature
Niemann	1968	4	Designation, Collapse, Initial, Equilibrium
Bormann and Likens	1979	4	Reorganization, Aggradation, Transition, Steady State
Oldeman	1989	5	Installation, Pre-ecuilibrium, Ecuilibrium, Elimination, Collapse (ecuilibrium defines Eco-unit equilibrium)

Different approaches were used because of the different nature of the forest mosaics for which the classifications were developed (Table 3.1). Phase definitions cannot always directly be related to the state of one regeneration unit. The decay phase and the regeneration phase, sometimes called Plenter phase (Leibundgut 1978b, 1982) used in the central European montane fir-spruce-beech forests are in fact complexes of small regeneration units, which are a typical feature of those sites (Oldeman 1989). Regeneration unit complexes are necessarily considered because mapping of separate, sometimes single tree regeneration units makes less sense. The phase definitions of Niemann (1968) and Bormann and Likens (1979) refer to the state of the sylvatic unit as a whole. The mosaic of regeneration units is characterized.

According to Leibundgut (1959) and Knapp (1974), stages, defined as developmental states within a succession series of different phytosociological associations, can be subdivided into phases, defined by Mayer (1976) as structurally clearly distinguishable developmental states within a forest association.

The terms stage and phase are used in a different sense by other authors. Hallé et al. (1978) interchange the terms. Korpel (1982) uses the term stage in a sense similar to Hallé et al., but uses phases as subdivision of the stages in a totally different way. Oldeman (1989), however, again conforms his phase definition to Leibundgut (1959). Problems with phase mapping arise as, for specific forest associations, margin values for criteria of delimitation, such as stand increment and volume (Mayer et al. 1979) are lacking (Henninger 1983). Without such values, the assignment of stands to developmental phases remains a procedure without much precision. As often in mapping, the method is limited by the precision of agreement on sometimes arbitrary criteria for delimitation and by map resolution (Koop 1981). Stands can be divided in the above-mentioned phases at different scale levels, resulting in a more or less coarse mosaic. In more detail, for example, a decay phase might be interpreted as patches belonging to a regeneration phase, a youth growth phase or a terminal phase.

Because borders of phases are partly drawn based on personal judgement, differences in interpretation will lead to different maps, but this is denied by Hallé, who experimented with different groups of persons mapping the same tropical forest in Sumatra (Oldeman personal communication). When the charting is repeated, e.g. as done by Korpel (1982), the interpretation of, and comparison with, the quite "soft" borders of former phase mapping is difficult, especially when these chartings are carried out by different persons.

Maurer et al. (1983) described computerized tree species mapping on digitized large- scale colour infrared aerial photographs. In a limited area, tree species are charted for groundtruth reference. Scale of mapping and groundtruth data collection of this method correspond closely with the nested plot design of this study.

Henninger (1983) and Bartelink (1988) evaluated possibilities of observations on sequential aerial photography to document series of forest development in circle plots in forest reserves. Circle plots are retraced on photographs taken at different points of time. Qualities for classification, derived from the photographs, are tested with groundtruth. Vooren and Offermans (1985) reported the use of large-scale aerial photographs, made from ultra-light aeroplanes, for recognition of individual senescent trees in tropical rainforests of the Ivory Coast. For the purpose of selective harvesting, senescent trees are detected just before the onset of their phase of decline.

Since air pollution damage to forest trees has been recognized over Western Europe, many authors have focussed on the possibilities of remote sensing and aerial photography for detection of tree damage, such as Kenneweg (1972 and 1983), Djawadi (1977), Mozgawa (1983), Rhody (1983), Kuhl (1984), Lustgraaf (1984), Spellmann (1984) and Masumi (1984).

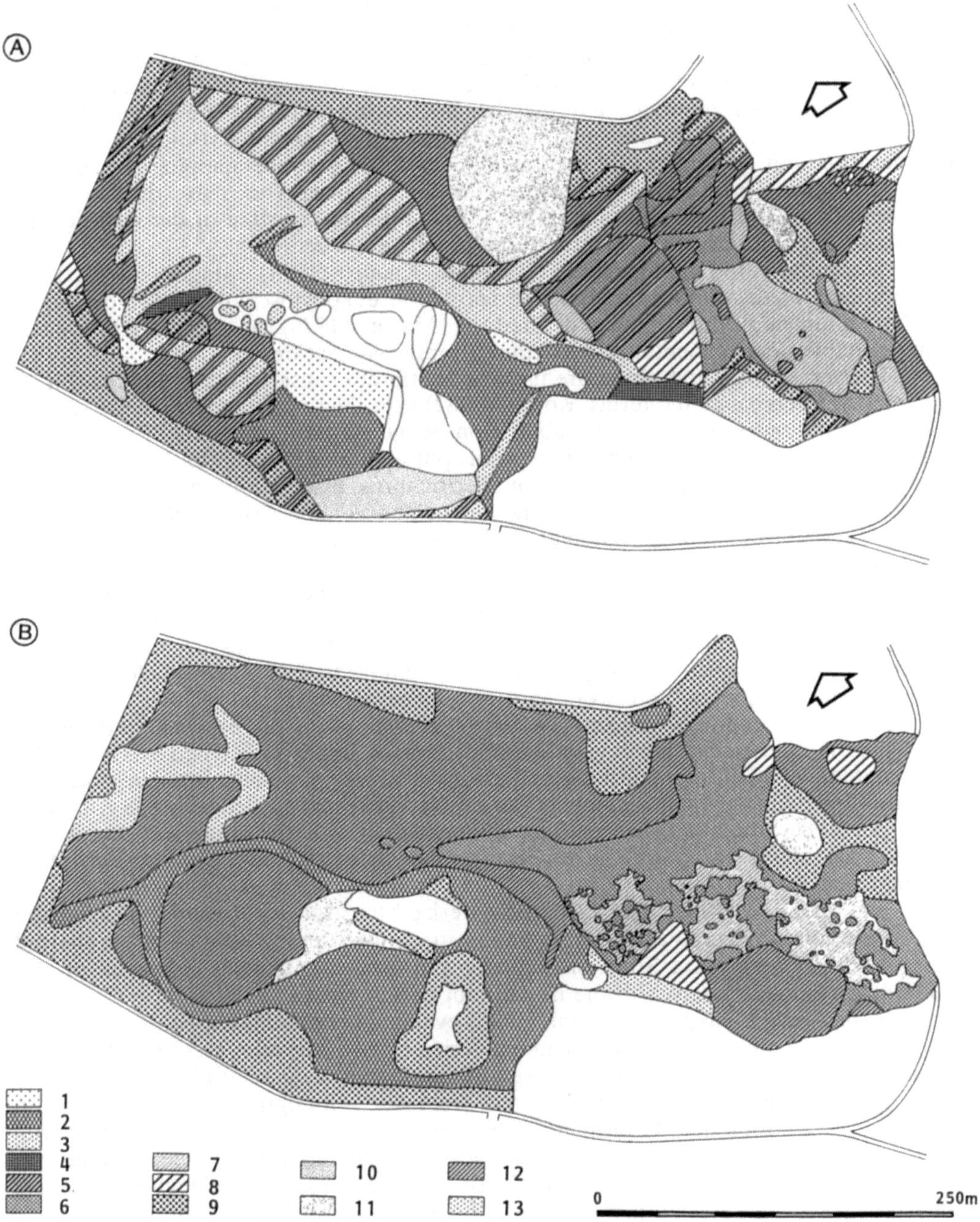
A
B
1
2
3
4
5
6
7
8
9
10
11
12
13
0
250m

Figure 3.2. Example of a repeated vegetation mapping of the forest type area of the Weenderbos (province Groningen, The Netherlands) based on repeated permanent plot relevés. **A** Vegetation map after Van Dijk and Westhoff (unpublished 1950). **B** Vegetation map by the author (1984). Remarkable shifts are the increase of the *Pteridium* type due to spread of canopy gaps. The permanent plots in the *Pteridium* type did not change at all since 1950. The decrease of the *Stellario-Carpinetum*, in 1950 described as *Querco-Carpinetum* associations, can be explained by the drop of the groundwater table after land improvement

In the developed monitoring system, aerial photography is used for mapping tree species- and phase distribution. For forest phase mapping at scale 1:2500, phases are used as distinguished in forestry according to metric height criteria. These height criteria are defined per site type as fractions of the maximum tree height. These phases can be mapped with aerial photography (Fig. 3.1b). The detailed three-dimensional groundtruth information collected in the strip transect and core area (see below) is used as a reference to interpret tree species distribution in the area beyond the core area (Chap. 7). In the ideal situation, a combination of simultaneous infrared black-and-white and true colour photography is used, to obtain best results in recognizing the tree species. The scale of photography is between 1:4000 and 1:8000. The most suitable scale depends on the forest type. The lower the forest and the smaller the tree crowns or the denser the canopy, the larger is the scale of photography required for optimal results. It is important that long-focus lenses are used to prevent distortion of the image in case of large-scale photography.

3.1.3 Mapping of Vegetation Composition

Vegetation units with characteristics that can be easily observed in the field are delineated on a map at scale 1:2500. Except where the area to be mapped is covered by topographical maps that show enough detail and pattern for guidance in the field, the aerial photograph also serves as the topographical base for mapping. Individual tree crowns and canopy gaps, that can be seen on aerial photographs, are used as references. Vegetation composition of the mapped units is sampled in fixed 10 x 10 m^2 plots.

The data on vegetation composition are noted on standard forms first or entered in computer files directly on a hand-held computer in the format: plot number - species code - cover - species code - cover - ...etc. These files are converted to a standard format: location number - date - plot number - species name - cover. Cover is put in different columns for herbs, mosses, species on dead wood, seedlings, saplings with a Dbh (=diameter at breast height) less than 5 cm. Trees are entered in three columns for height classes that can be defined separately for each sample. This format in turn is loaded in the database (see Appendix). Vegetation types are established after processing the samples with the help of cluster techniques (e.g. Hill 1979a,b; Jongman et al. 1987), sometimes followed by final classification by hand. Local vegetation types are thereby distinguished. For the situation in The Netherlands these local types are assigned to a Potential Natural Vegetation type (Van der Werf 1989). With the help of the local vegetation types distinguished, the legend for the mapping units is established.

Depending on the coarseness of the vegetation pattern with regard to the detail permitted by the resolution of the map, the units can be identical to, or cover complexes of, local vegetation types. Figure 3.2 shows the results of a repeated vegetation mapping.

Some major vegetation shifts can be identified, although the legends of both maps are not identical. The content of the old legend was documented with 18 vegetation plots, accurately indicated on the map. In 1985 these plots were reassessed.

3.2 Core Area

The core area covers at least some regeneration units. At this scale level the interior organization of the regeneration units is described in terms of their subsystems, the woody and non-woody compartments. Trees are charted, individually measured and classified on the basis of quantitative values for the different variables. The pattern in the non-woody compartment is mapped.

To allow explanation of the effects of trees upon the herbaceous vegetation in the herbaceous transect area, information on forest structure over a surface ranging at least one maximum tree-height around the herbaceous transect area is required. In the case of tree heights between 20 and 30 m, the core area therefore measures 70 x 140 m (0.98 ha). In the case of forests with tree heights less than 20 m the core area width can be reduced to 50 m, with a length of 140 m, resulting in an area of 0.7 ha.
Values of variables, collected in the core area for trees with a Dbh more than 5 cm, can be divided into variables related to forest structure (tree chartings), quantitative tree measures and estimated tree classifications. In addition, the core area is documented with sequential ground photograpy and its forest floor vegetation is described by mapping. Trees and shrubs with Dbh less than 5 cm are assessed in the herbaceous transect area.

3.2.1 Tree Charting

Regeneration units and stand structure have been studied by many authors with the help of profile and ground-plan drawings. Especially when dealing with complex forest structures of selection forests, scientists have used methods of profile drawing, e.g. Blomqvist (1879), Ammon (1937) and Knüchel (1944). For the purpose of transfer of knowledge on selection or for thinning, the method was used by Köstler (1953), Koop (1981) and De Graaf (1986). Oldeman and Schmidt (1986) use it for silvicultural design ,for the time being, for educational purposes only. The profile drawing method has also frequently been used for the description of virgin forests or spontaneous forest development, e.g. by Watt (1923), Fröhlich (1930), Mauve (1931), Davis and Richards (1933), Leibundgut (1945), Mayer-Wegelin et al. (1952), Richards (1952), Oldeman (1972, 1974a/b), Whitmore (1975), Hallé et al. (1978), Lamprecht (1980) and Koop (1981). Rollet (1974) and Oldeman (1974a/b, 1989) give many examples of transect drawings.

Forests are too complex for all properties to be predicted, and every method of transect drawing is therefore an abstraction, adapted to the question asked (Oldeman 1986). Line and strip transects have their restrictions in being representative for a larger surface area and efforts have been made to obtain a more spatial three-dimensional image of forest structure. N. Hallé, who illustrated an article by Aubréville (1965), tried to depict forest structure by means of three-dimensional block diagrams. Although Hallé drew all African forest types, Richards (1983) holds the method useful only for relatively open forests such as savannah forest. The images give an overall characterization of forest structure. Because of obstruction by too many trees, their information cannot be compared with the details of forest structure that can be depicted in a transect of appropriate width. Some authors drew adjacent transects (Schauer and Stöcker 1976; Mayer and Neumann 1981). As has been emphasized by Rollet (1974), the representativity of one single transect for the total of the forest structure of a stand is low.

Extensive chartings of stem base positions and dead wood on maps of areas up to 1 ha were made by Watt (1923), Markgraf and Dengler (1931) and Falinski (1977). Genssler (1979), Lamprecht (1980) and Bücking et al. (1985) describe the so-called 1 ha "Kernfläche" charted in natural forest research projects in several states of the F.R.G. Prusa (1985) published maps of all trees with a diameter at breast height exceeding 10 cm over entire forest reserves with sizes of up to tens of hectares. In Baden-Würtemberg besides the core area, circle plots of 40 m diameter in a 1 ha grid over the whole forest reserve are mapped (Bücking et al. 1986). Bouchon et al. (1973) drew maps of crown projections only, over an area of 35 ha in France. The crown projections were assigned to and distinguished after different diameter classes of the tree bole. Ohsawa (1981), Nakashizuka and Numata (1982), Nakashizuka (1985), Naka (1982), Leemans (1986) and Torquebiau (1985) charted tree base positions, dead wood and crown projections.

Only few authors charted microrelief due to uprooting. Sometimes the position of lying dead wood is drawn. Uprooting is a common forest process. Many authors reported on it, especially in regions with remnants of virgin forests or where original forest floor micro-relief had not been levelled (Lutz 1940; Stephens 1956; Armson and Fessenden 1973; Stone 1975; Brown 1976; Koop 1981; Beatty 1981). Brewer and Merritt (1978) and Hutnik (1952), Lyford and Maclean (1966) and Putz (1983) reported on the influence of uprootings upon tree distribution. Londo (1977), Beatty (1981) and Koop (1981) described the influence on species composition in the herb layer. Oliver and Stephens (1977) and Zeide (1981) describe methods of mound dating by research on trees that grow on them.

Henry and Swan (1974) used the information provided by the position and state of decay of dead wood for reconstructing the history of a forest on a restricted area. Falinski (1977, 1986) showed that repeated charting of dead wood can be useful in detailed studies on dead wood dynamics. Mortality, translocation and decomposition of wood, as an important part of the forest ecosystem flow of organic material, has been studied by Falinski (1977) and Sollins (1982). Yoneda (1975) described properties of decaying wood.

Brown (1974) and Martin (1976) describe line intersect methods for the quantitative inventory of fallen woody material. The method can be used to estimate biomass over large areas.

In the monitoring system, positions of tree bases of dead and living trees as well as those of the stumps and uprootings are charted every 10 years. Abrupt changes such as tree uprootings and breaks are charted every 3 years, simultaneously with the herbaceous transect area description. Crown projections are drawn with the help of imaginary vertical planes that touch the crown. A sketch of the intersecting lines of these imaginary planes with the ground plane providing a polygon in which the crown projection fits (Fig. 3.3). Devices such as the "Cajanus tube" enable an exact vertical view angle for assessment of the crown projection. Such devices may be used successfully in open forest stands. Touching or overlapping tree crowns can only be distinguished separately with great difficulty. In these cases the unaided eye, that can judge distances and relative positions, is the best to assess the limits of crowns. Crown projections of all trees, with their tree base within the core area, are completely drawn, within as well as outside the core area limits. For identification all trees are numbered on the maps.

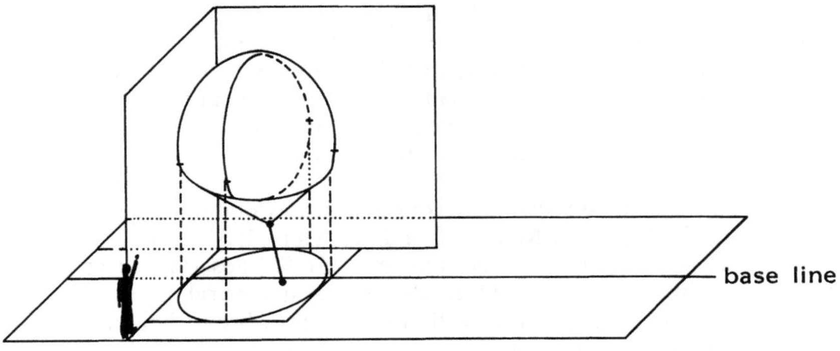

Figure 3.3. Crown projections can be drawn using imaginary vertical planes that touch the tree crown. The lines of intersection with the ground surface make up a polygon in which is wrapped the crown projection to be drawn. The vertical planes can best be imagined parallel to and at right angles with the base line of the transect respectively

The hand-drawn chart of the core area serves as the first control in checking for mistakes. For three-dimensional model processing (see Chap. 4.3) only measurements of the x- and y-co-ordinates of tree base and crown extremities are neccesary. Notes on standard forms of these measurements, without the visual check of a tree drawing, facilitates mistakes, e.g. in the sequence of tree co-ordinates or in interchanging of tree numbers (Schulting and Staaveren 1981). Direct input of field-measured co-ordinate values in a hand-held computer, with an automatic check for impossible combinations and a visual check provided by a small graphical display of the on-going charting on the computer screen (Fig. 3.4), has proved to be an acceptable alternative for drawing (see Appendix). Mistakes and confusion about tree numbers can be excluded by a built-in error tracer, while there is also the possibility of direct correction in the field.

The tree positions, crown projections, positions of fallen dead trees and uprootings, drawn to scale 1:200 on millimeter paper sheets in the field, are transferred in ink upon shrink-proof tracing sheets. Copies of the sheets with only the tree positions are used to make a tree number chart, a tree species chart and a blank tree position chart to be used when the plot is to be charted again. These charts and two duplicates are safely stored in separate places. The complete chart with information on tree positions and crown projections is superimposed on the tree number chart and put on a digitizer tablet. For sequential tree numbers, the tree base and peripheral points of the crown, necessary for three-dimensional data processing (see Chap. 4.3), are digitized with the help of the computer program DIGIT (see Appendix). This program enables automatic number increment and easy program jumps for correction of arbitrary tree numbers.

In a drawing, without loss of time, supplementary information is automatically included. This information is lost if field-data on forest structure are collected only with a hand-held computer. However, for the extensive core area charting, this loss balances the gain of time for working out the field drawing, transferring it to tracing sheets, followed by digitizing the co-ordinates. In addition, besides computerized inventory, part of the core area in the woody transect area is drawn by hand.

Fig. 3.4. Hand-held computer for the input in the field of forest structure measurements. **A** A small graphical display provides a visual check of the on-going charting. **B** A tree caliper can be connected for tree diameter measurements. **C** A tree altimeter can be connected for tree height measurements.

3.2.2 Tree Measurements

Figure 3.5 shows all height measurements for a tree. Moreover, diameter is measured at breast height (Dbh). Heights are measured for the crown top (Ht), the greatest width of the crown or the periphery height (Hp), the crown base (Hc), the first living fork (Hf), and the relative height of the tree base to a reference height (HB). In the case of bent stems or where stems overtop a partly dead tree crown, and when the height of the top of the living tree crown (Ht) is less than the stem length, the length of the stem is recorded.

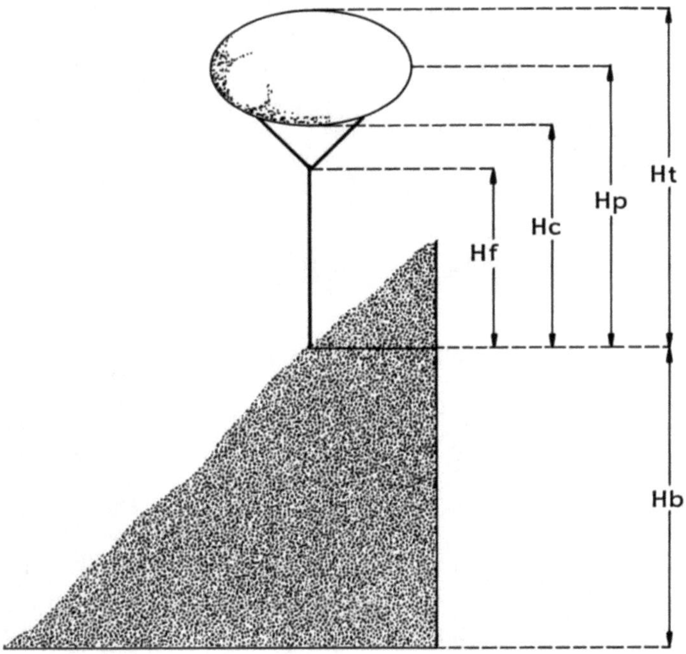

Fig. 3.5. Height measurements on a tree, *Ht* (top height), *Hp* (height of the greatest width of the crown the periphery), *Hc* (height of the crown base), *Hf* (height of the first living fork), *Hb* (height of the tree base to a reference height).

3.2.3 Tree Classifications

In the already mentioned 1 ha "Kernfläche" of the natural forest research projects in several states of the F.R.G. (Genssler 1979; Lamprecht 1980 and Bücking et al. 1985) besides charting individual trees, often individual trees have been measured and classified according to Leibundgut (1959, 1982) and Mayer (1976).

Fig. 3.6. Photographs of dead wood in different stages of decay. **A** Fresh fallen tree with intact bark (class 1). **B** Tree with decayed bark (class 2). **C** Superficial decayed tree with soft wood that can easily be squashed up to a few centimeters (class 3). **D** Thoroughly decayed wood, that has lost its consistency. **E** Remains of the mouldered tree are to be found in the litter layer and very often indicated by *Oxalis acetosella*

A
B
C
D
E

Oldeman (1974a) uses architectural codes for separate trees, that indicate the so-called potential trees, trees of the present and trees of the past. According to Hallé et al. (1978), trees of the present are trees that lack potential for further crown expansion. They exist at a quasi-stationary height, without adding to this dimension or to the width of their crowns. Trees of the present generally determine or dominate the architecture of the forest they live in. They can do so during long periods, provided the forest is not subject to frequent and violent total destruction (e.g. by hurricanes). No further "promotion" to a "higher" rank is possible. In contrast, a potential tree (formerly called "tree of the future" and later renamed to prevent confusion with an existing forestry term) is a tree that either is in a phase of vigorous expansion, or suppressed and waiting for conditions which will promote resumed growth. These trees still have a potential for future expansion. Trees of the past are trees that are damaged beyond repair so that there is no potential for future development, nor a possibility of survival in a stationary state. Such growth potential estimates of forest trees are often included in the transect by drawing trees in different line symbols. In this way the dynamics (past, present and future) of the forest plot are indicated. In the monitoring system, the trees are classified in the field as to their species, architectural code according to Oldeman (1974a), vitality, and cause and intensity of damage (Leibundgut 1959, 1982; Mayer 1976).

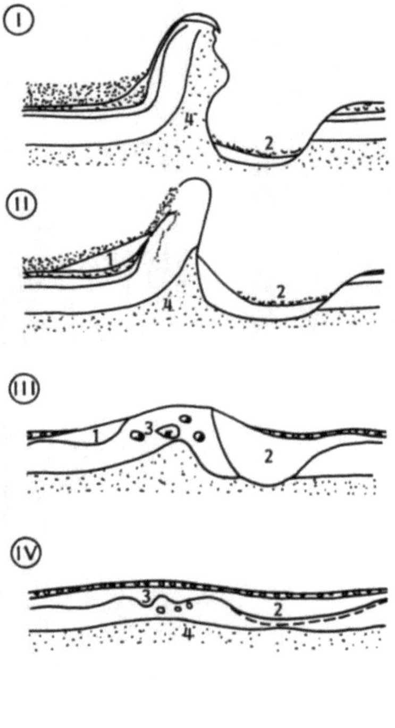

Fig. 3.7. The process of flattening of the microrelief due to uprooting of a tree can be divided into four stages (Roman numbers) after Koop (1981). *1* Soil material formerly situated above the main root layer. *2* Soil material formerly situated underneath the main root layer, intermingled with organic material having accumulated in the pit. *3* Loose soil mass with mouldered roots and many cavities. *4* Deeper uplifted soil layer

The cover of the crown within its projection is estimated for the individual trees with the decimal scale as published by Doing-Kraft (1954) and Londo (1984) (see Appendix). The position and stage of decay of dead wood is recorded as shown in Figure 3.6. Stage of flattening of microrelief caused by uprooting, the height of the mound and depth of the pit are noted according to Koop (1981) (Fig. 3.7).

A photocopy of the hand-drawn tree number chart or, in the case of computerized co-ordinate input in the field, a computer-plotted tree number chart (Chap. 6) is taken to the field for reference to take measurments per tree. The measures and classifications are entered on standard forms or again directly entered in the hand-held computer.

The data on the standard forms and the digitized plan are loaded in the computer database with help of the program ORACLIN (see Appendix). The combinations of data from the standard forms and the digitized plans are first checked for errors or impossible combinations for each tree number with the program CORRECT (see Appendix). Examples of the last categories are: missing (i.e. not digitized) crown projections of live or dead trees that do have a tree crown projection. The sequence of digitized peripheral points is checked and for a visual check forest structure is plotted (Chap. 6). Data collected with the hand-held computer are already corrected in the field and can be loaded directly.

3.2.4 Tree Ring Sampling

For the core area, or a central part of it, tree ring samples are collected for all trees exceeding a Dbh of 10 cm. These trees are sampled twice at breast height with a 40-cm Pressler borer in opposite directions where the diameter is the same as the mean diameter deducted from girth measurement. The samples are stored in tubes filled with a 1:2 solution of glycerin and methylated spirit. The bore holes in the trees are coped with a disinfectant paste (Kopinga 1979). Tree ring samples are measured with a linear digitizer (Van der Beek and Maessen 1981). Tree rings hit askew, for example near the core of the tree, were repositioned at right angles with the tree ring. The values of cumulative tree ring width are directly written in the files. The program CORRING (see Appendix) plots tree ring width of several trees on the screen or sends it to the printer. To achieve absolute dating of tree growth the thus printed dips and tops in the sequence of tree rings of individual trees have to be collated and related to the sequences of years of documented extreme weather conditions, e.g. dry summers, cold springs or years with heavy storms (Fritts 1976). Between two dated tree rings the number of tree rings has to be fitted. Input errors can be traced, and possibly misinterpreted tree rings, which were marked during counting, have to be divided extra or lumped together. From the two samples of each tree, mean values of annual tree ring width are computed.

3.2.5 Sequential Ground Photography

Drawings are analytical abstractions. Photographs cannot replace profile drawings, tree charts or vegetation maps. During drawing, scaling is used more or less consciously for depiction. Details at lower integration levels are left out, for example the tree crown is drawn as a transparent body without leaves. A ground photograph always contains noise and information which is of less importance, e.g. it suffers from obstruction caused by nearby trees or shrubs.

Aerial photographs do not show the lower tree crowns, hidden by higher canopy trees. Thalen (1980) reviewed the possibilities of the use of repeated aerial photography and sequential ground photography in permanent plot studies in a variety of vegetation types. He concluded that the combination of conventional aerial photography (vertical stereo at scales 1:5000 to 1:50,000), large-scale 70 mm photography (stereo, colour or colour infrared at scales 1:500 to 1:5000) and ground photography together with permanent plots seems to be the most promising approach. Bücking (1986) reported on a method of sequential ground photography in combination with permanent sample plots. About 16 partly overlapping photographs (f = 50 mm), covering a view angle of 360°, are taken from one point in the center of a sample plot. De Graaf (1986) used stereo pairs as a document for depicting selective silviculture in tropical rainforest in Surinam. The problem of sequential photography at marked camera points is that the previously chosen view angle may become obstructed during vegetation development or the camera point may become unattainable or lost.

In the monitoring system at fixed and marked camera points in the core area, every 5 or 10 years, depending on the rate of change of the particular forest vegetation, black and white photographs (negatives 60 x 60 mm) and colour-slides (24 x 36 mm) are taken. A 28-mm wide-angle lens is used and camera-stand height is 1.75 m. At a certain point two overlapping view angles are chosen, one obliquely downward to record the ground vegetation and one obliquely upward for the forest tree structure. The position and viewing angle of a camera point is drawn on the 1:200 maps of the core area (Fig. 3.8). Camerapoints are marked on the tree chart of the woody transect area and core area. The points are digitized in x and y values and stored in the database together with the name of the author, date, serial number of the camerapoint and record number of

Fig. 3.8. Example of ground photography with two overlapping view angles. **A** One obliquely upward to record forest structure. **B** One obliquely downward for ground vegetation in the Otterskooi (province Overijssel, The Netherlands)

the black and white negative and colour slide. Prints of the black and white photographs together with the colour slides are kept separately from the black and white negatives. Black and white negatives can be kept longer without quality loss than colour slides. Therefore when for some reason only a set of colour slides is available, black and white reproductions are made.

3.2.6 Vegetation Mapping

A main restriction in a considerable number of cases of transect analysis and tree charting is that only trees of a diameter at breast height over a certain limit are shown. No other forest components are incorporated, except for large lianas or tall herbs that can be depicted at the scale used. Sometimes a complete inventory or results of sampling of smaller species, that cannot be depicted, is added. Examples of combined inventory of complete species composition in adjacent plots and transect analyses are given by Walter (1979), Bachacou and Chessel (1979), Barthélemy (1979) and Koop (1981). Londo (1971, 1978) discussed the role of repeated vegetation mapping in addition to permanent plot survey to obtain maximum information on the relation between structure and dynamics of vegetation types. Remapping can detect shifts in vegetation pattern, but changes in the vegetation composition are only detectable through detailed inventories of repeated series of relatively small permanent plots, chosen preferably in a grid or in transects in gradient situations to represent various aspects of the vegetation pattern, mapped over a large area.

In the monitoring system a detailed vegetation map of the core area is made at a scale of 1:200. Vegetation units that can be easily seen by virtue of distinct characteristics are distinguished in the field. Depending on the variation of the species composition in the core area, differences between these vegetation units may vary. Such units can be characterized by a clearly different vegetation structure or dominance of one or a few species. Only in exceptional cases have these vegetation units a completely different species composition at the level of a different site. A legend for characterization of the mapped vegetation units is established by processing species composition of the herbaceous transect plots with cluster techniques (Hill 1979a,b; Jongman et al. 1987), and when necessary with final classification by hand. In case mapped vegetation units are not represented in the herbaceous transect plots, additional 10 x 10 m samples or smaller vegetation plots are assessed and used for the classification as well. These extra samples are fixed on the core area map with tree base positions, to be retraceable for repeated description. The vegetation mapping helps to generalize conclusions on vegetation development as observed in permanent quadrats. On the other hand, the 1:200 map gives more depth to the information content of the 1:2500 vegetation map.

An example of both vegetation maps is presented in Figure 3.9. The overall sequence of vegetation types on the large-scale map 1:200 corresponds with the small-scale map 1:2500. Because of the large scale (1:200) of the vegetation map of the core area, pure legend units could be shown. There was no need to distinguish and include at this scale (legend) units made up of complexes of more than one vegetation type. In fact, some extra variants of the vegetation types described for the whole reserve could be distinguished and their distribution could be mapped in more detail. On the other hand, some of the vegetation types are found within the core area in a form poorer in species. From the northeast to the southwest the core area is divided into the *Phragmites* variant of the *Dryopteris* type, the typical variant of the *Dryopteris* type and the *Impatiens* type.

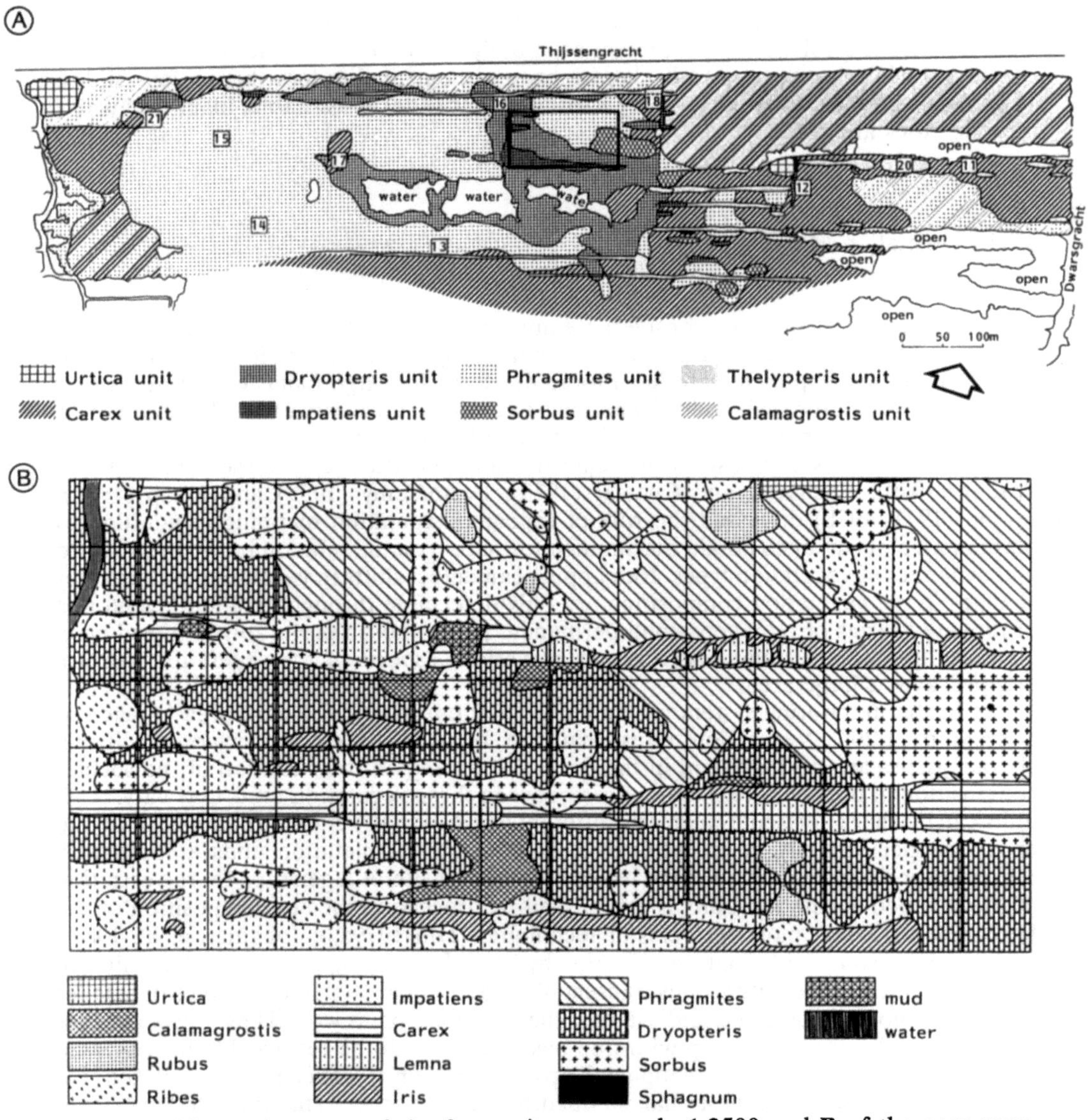

Fig. 3.9. A Vegetation map of the forest site area scale 1:2500 and **B** of the core area scale 1:200 of the Otterskooi (province Overijssel, The Netherlands). *B* indicated in *A*

Many small isolated patches of the *Sorbus* type are found on the highest parts and on the higher edges of the ditches that originate from the spoil dredged from the ditches. Scattered small patches of the *Impatiens* type are found in depressions within all vegetation types. This large-scale vegetation map illustrates adequately the nature of the legend units of the small-scale map. All vegetation legend units of the small-scale map are clearly shown to be vegetation complexes of more than one type in a fine mosaic mainly caused by relief of the pools and ditches of peat cutting (Table 3.2).

Table 3.2. Distribution of the vegetation types (horizontally) over the legend units (vertically) of the 1:2500 vegetation map of the Otterskooi (see Chap. 8). Dominant or co-dominant types are marked with an asterix (*) and subordinated types with the plus sign (+)

	Sorbus	*Dryopt.*	*Phrag.*	*Calmag.*	*Carex*	*Impat.*	*Thelyp*
Sorbus	*						
Dryopteris	+	*				+	
Phragmites	+	+	*	+	+	+	
Calamagrostris	+		*	*		+	
Carex					*	+	
Impatiens						*	
Thelypteris						+	*
Thelypteris/Carex		+		*	+	*	

3.3 Woody Transect Area

For the study of subsystems constituting the "skeleton" of a regeneration unit, i.e. the component trees, a metric standard plot size is chosen, large enough to allow analysis of at least one regeneration unit within its borderlines.

3.3.1 Profile Drawings

For delimitation of the trees to be included in the transect, the line transect method (Oldeman 1980; De Coo 1981) is used besides the strip transect method. In line transects all trees are drawn with a crown of which the projection intersects or touches the midline of the transect. The advantage of this method is that, to keep the drawing readable, no attention has to be paid to an appropriate transect width, whether traversing a dense or an open vegetation. On the other hand, however, no translation of densities to surface area can be made. The method can be used for selecting transect intervals in an "objective" manner for more detailed analyses.

In a strip transect all trees that have their tree base within the borderlines of a rectangular strip are drawn. The optimal width of the strip transect depends on stand density, in the way that optimal thickness of the microtome slices depends on the nature and size of the cells in a tissue. According to Oldeman (1980), in general a width

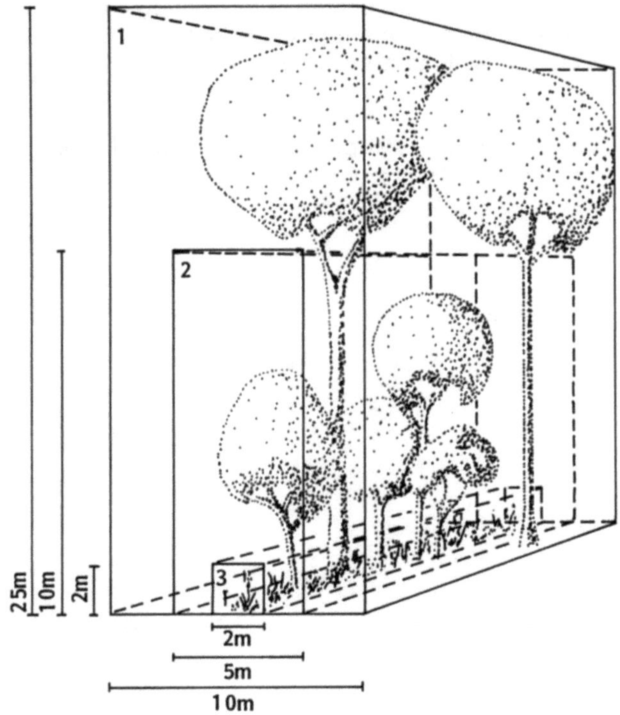

Fig. 3.10. Procedure of transect drawing according to Koop (1981). In order to keep a side view readable, trees of decreasing size must be depicted in proportionally narrower strips (*2*, *3*) within the woody transect area (*1*) in which the highest trees are drawn

between one half and one third of the tree height gives the best results. To preclude that profile drawings with dense shrub layers, drawn over the total width of the transect plan, become unreadable, tall trees can be drawn in a wider transect than lower trees and shrubs. Some authors leave parts of the crowns, obscured by other trees' crowns in front, out of the profile drawing. Others draw all trees in a schematic transparent style, which has the advantage over photography or the more "realistic" style of drawing of an unobstructed transparent slice of forest structure. For reasons of readability, tree stems obscured by other stems in front are moved half a metre in the drawing. The real position of the tree base can be found in the ground plan. In the legend, attention has to be drawn to this deviation. Sometimes trees outside the transect that have an impact on the transect structure are drawn as well. Koop (1981) also drew tree crowns bordering canopy gaps that were traversed by the transect. If a transect ground plan is added it gives additional information, e.g. on the extension and orientation of gaps that might pass oblique radiation into the transect.

A ground plan of the profile strip is not always drawn as, according to Richards (1983), it ought to be done. Often the width of the strip transect is not stated. When the plan is drawn, some authors draw only the positions of the tree base. If the crown projection is added, the perimeter may be drawn only within the boundaries of the transect (Oldeman 1974a) or the complete crown projections of all trees that have their stem within the transect are depicted. In the latter case, crown surface for an individual tree can be derived.

The scale used and detail shown in the architecture of trees and other forest components depends on the aim of the transect study. When the architecture of the individual tree or the detailed structure of a stand at regeneration unit level is to be studied, larger scales for a restricted area of some 100 m^2 are used and ample attention should be paid to branching pattern, windings and remnants of previous forks on the trunk. Oldeman (1980) for this purpose makes separate sketches of each tree in the field, which later on the desk are collated with the help of positions of the tree bases. At the crown periphery inherited reiterative patterns of tree growth are included consciously in lieu of the reality that might have been seen in the field (Oldeman 1974a). When less detail on an individual tree is needed, profiles can be drawn on a smaller scale. All trees can be drawn to scale, both in profile and in vertical projection, on one piece of paper (Koop 1981). This method has the advantage of immediate control in the field of the relative positions of the tree crowns. A possibility to split the transect into two parallel transects of half-width leaves more room for detail in both separate profile drawings (Van de Winckel 1980). When transects are made for floristic or physiognomic purposes or for the study of the forest mosaic as a whole (sylvatic unit level), a small scale is used, and trees are drawn schematically. Crown circumferences are drawn by Whitmore (1975) and Brünig (1977), without an indication of branching patterns or architectural code. Among others, Mayer et al. (1979), Zukrigl (1970) and Korpel (1982) use formalized images for different tree species and pay little attention to variation in individual branching pattern. This style mainly developed in the central European mountainous fir-spruce-beech forests. Information on how the transects were drawn, what points were measured and what was sketched or estimated here is often lacking.

In the monitoring system the tree layer of a permanent transect measuring 10 x 100 m is characterized, in sufficient detail to describe tree shape, with profile drawings and crown projections at scale 1:200 at 10 year intervals. To keep profile drawings readable, lower and denser vegetation strata of the forest are drawn for a narrower strip within the 10 x 100 m transect (Fig. 3.10). Heights over 10 m are measured, others are estimated using 4 m jalons as a reference. Crown circumference is drawn and the stem and most important branches are sketched (Fig. 3.11). Detailed branching patterns are not included in view of the time required in the field, the intended storage and retrieval system, and last but not least the intended range of applications of the system.

3.4 Herbaceous Transect Area

The herbaceous transect area, split up into adjacent 2 x 2 m quadrats, enables analysis of species composition within the map units of the 1:200 vegetation map of the core area. The complete 200 m^2 surface of the 50 quadrats covers the minimum area of a forest vegetation sample according to Mueller-Dombois and Ellenberg (1974).

3.4.1 Permanent Quadrats

To obtain information on the relation between structure and dynamics of vegetation types, Londo (1971, 1978, 1982) discusses the role of repeated vegetation

Fig. 3.11. Examples of the profile drawing of the transect area of Fontainebleau I (France) and Leuvenumse bos (province Gelderland, The Netherlands)

mapping in addition to permanent plot survey. Glavac (1972) suggests for the analysis of herbaceous vegetation development in natural forests many small plots of 1 to 4 m^2 arranged in a certain pattern, in a given case a transect. Houllier (1985) argues for sampling with partial replacement with at least a 50 % reassessment in permanent plots and another 50 % of temporary plots. Falinski (1977) uses adjacent permanent plots of 2 x 2 m in a 50-m transect that is situated in a 1 ha block where tree canopy structure is vertically photographed in a grid of 10 x 10 m and where tree position and dimensions are measured. The plots are photographed weekly, using a camera-stand that has a unique fitting on the metal framework that outlines the plots. The long-term photographic series can afterwards be used for different purposes and analyzed in the laboratory. An advantage of smaller plots is that they are easier and more accurate, and thus repeatable cover assesment.

Since Braun-Blanquet (1964) developed the first scale for cover assessment, many modifications have been proposed (Knapp 1984). Londo (1984) reviews the development and points out the advantages of the decimal scale for relevés of permanent quadrats, first suggested by Doing Kraft (1954). Sykes et al. (1983) indicate mistakes that can be made with the use of visual cover assessment as quantitative estimators of some woodland taxa. Some of the sources of bias can be minimized by assessment by the same person. A limited plot size raises accuracy in cover assesment, but with decreasing plot size, the bias resulting from the edge effect is increased.

In the monitoring system, herbaceous vegetation, including mosses, liverworts, tree seedlings and saplings and shrubs with a diameter of less than 5 cm, is described in 50 adjacent 2 x 2 m plots along the centre line of the transect every 3 years, using the decimal scale for relevés of permanent quadrats (Londo 1984). Herbaceous transect areas of study areas abroad are for logistic reasons described every 5 years, e.g. Bialowieza (Poland) 1977-1982-1987, Fontainebleau (France) 1981-1986. The permanent quadrats describe the interior state of the pattern units of the non-woody compartment. Vegetation composition of adjacent permanent quadrats in a transect, traversing the vegetation pattern,is used to describe the map units and the shifts in vegetation pattern in the core area. Vegetation data for the 50 permanent plots in the herbaceous transect area are entered either on standard forms or directly in files on the hand-held computer. Data from standard forms are later entered in files according to the format: species code followed by a sequence of plot numbers each followed by the cover of the species in that plot: species name - plot number - cover - plot number - cover - ...etc. (see Appendix). Similar to the 10 x 10 m plots in the forest type area these files are converted to a standard format and loaded in the database (see Appendix).

Part B Data Processing

Chapter 4

The Tree Model

To overcome the limitations of the two-dimensional images of plan and profile drawings, a three-dimensional forest model has been developed that plays a central role in the data processing and modelling. First, literature dealing with tree models relevant for the forest model development will be reviewed. This is followed by a description of the tree component model developed for the information system.

4.1 Review of Tree Models

4.1.1 Models of Tree Architecture

Hallé and Oldeman (1970) and Hallé et al. (1978) define the architecture of a tree as "the visible, morphological expression of the genetic blue-print of a tree at any one time". Based on analyses of tree growth with the help of the architectures of many observed trees, the authors developed the abstract concept of the architectural model. The model can be seen as the plan of growth of a tree. The tree architecture is the skeleton that fills the ecotope, i.e. the space the tree fills during its life-span (Oldeman 1974b). References on structural features of individual trees can be found in Hallé and Oldeman (1970) and Zimmermann and Brown (1971).

Different computerized models have been developed to simulate tree architecture using the principles laid down in the first hand-drawn tree models. Deterministic models consistently repeating the branching pattern illustrate the effect of changing values for the parameters of branching on form (Honda 1971), or on form during growth (Edelin 1977, 1984; De Reffye 1979; De Reffye and Edelin 1989). Fisher (1984) argues that even if a model includes stochastic probability in development of meristems that may closely fit the pooled real distribution of a tree or population, still each individual meristem and module (branch unit) is responding to a wide variety of exogenous and endogenous influences, each presumably with its own probability of plant response. Reiteration is per definition unpredictable (Oldeman 1974a) and occurs apart from the programmed development, especially in species in stressed climates or circumstances. Although such computer models of branching pattern and crown geometry may produce realistic results of simulations, there is a fundamental problem in using these to infer adaptive tree growth. Because influences from outside, for example internal and external shading, are not used for prediction of meristem responses and branch extension, the summarized growth effect cannot explain shape or development of an actual tree inserted in a stand. Tree shape can only in general terms be correlated with its architecture as done by Hallé et al. (1978), describing physiognomic tree shapes as simple overall stereometric surfaces of revolution, e.g. funnels, spheres, hemispheres and discs.

4.1.2 Two-Dimensional Tree Shape Models

For two-dimensional computer-plotted depiction of measured trees in profile section Brünig (1977) and Werger et al. (1984) used rectangular crown shapes on perpendicular stems. Horn (1971) uses three parameters to describe the wide variety of shapes of tree crowns in transverse section: absolute size (c), ratio of height to width (b) and degree of convexity (a) in the Eq. (1).

$$x^a + (by)^a = c^a. \tag{1}$$

X and y are variable Cartesian coordinates. When a = 1 the equation represents a straight line, when a > 1 it becomes convex. This equation behaves nicely in the quadrant where x and y are positive. For the other quadrants the equation of the positive quadrant must be reflected. Cheviron (1985) uses a similar approach.

For depicting trees in computer-plotted profile drawings, Kätzler (1984) and Bücking et al. (1986) presented a model for broadleaved trees constructed of four quarters of ellipses: two quarter ellipses above the greatest crown width, that is fixed at 1/4 of the crown length and two below this width. For white fir (*Abies alba*), the greatest width is fixed at three-quarters of the crown length. Norway spruce (*Picea abies*) crowns are represented by overlapping triangles. Inclination of the stem is derived from the relation between the projection of the tree top and the tree position on the ground. Crown projections, measured at six extreme points with compass-angle and distance to the tree base, are depicted as a polygon. Although these models meet with the need of visual expression they are two-dimensional and not suitable for explanatory forest modelling applications.

4.1.3 Three-Dimensional Tree Shape Models

Many rather simple surfaces of revolution have been used for tree shapes in forest simulation models. Cones are used by Ford (1976), Cannell (1983), Hari et al. (1985) and Mohren (1987). Mohren used cones of different height/diameter ratios. Satterlund (1983) modelled trees as a crown consisting of a cone at the top of a cylinder and the bole as a solid cylinder. Akça (1983) and von Spellmann (1986) present a tree crown model to be measured in aerial photographs from four points on the visible periphery of the crown and a fifth point representing the top of the tree. The visible part of the tree, called "Lichtkrone", is calculated from a surface of revolution that takes into account the individual crown form. Needle-leaved trees, such as pine (*Pinus sylvestris*) and Norway spruce (*Picea abies*), have an average crown form between a cone and a quadratic paraboloid. Broad-leaved trees, such as beech (*Fagus sylvatica*) and oak (*Quercus robur* and *Q. petraea*), come close to a quadratic paraboloid and a hemisphere. For simulation of carbon flow in *Pinus radiata* stands, Rook et al. (1985) used an ellipsoid crown shape. The above mentioned cone-shaped models for coniferous trees and the elliptic shapes for broadleaved trees come closest to the requirements of the system presented in this study (Chap. 1). However, because they imply a symmetric surface of revolution, positioned by a vertical rotation axis which is the bole of the tree, they simplify asymmetric crowns and bending trees, which are common features in many natural and seminatural forests.

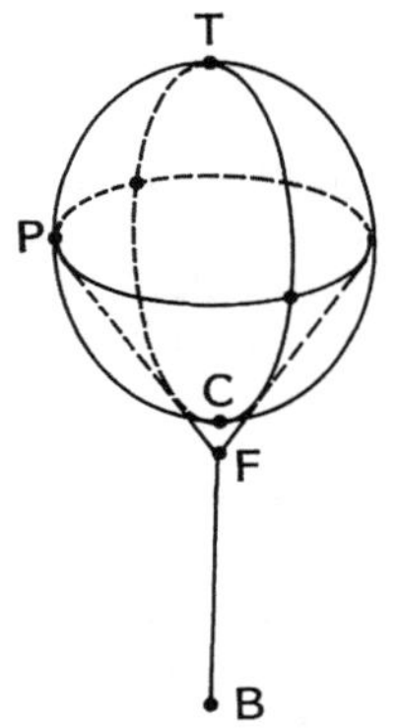

Fig. 4.1. The three-dimensional tree model. *T* top of the tree; *P* periphery point; *C* crown base; *F* fork; *B* stembase

4.2 Tree Model Description

For computerized data processing, the volume occupied by one tree at a certain moment in a certain forest stand is described by the co-ordinates of eight points, representing: tree base, place of first fork, top and base of the tree crown and the four extremities (peripheral points) of the crown circumference in positive and negative x- and y-direction. All z-co-ordinates are derived from height measurements while x- and y-co-ordinates are digitized from the hand-drawn plan. X- and y-co-ordinates of the top and base of the crown and of the fork, that are not drawn in projection are computed separately. By making use of the ellipsoid nature of most deciduous tree crowns, a best-fitting wrapping can be constructed by connecting the four peripheral points of the crown circumference with quarters of ellipses, using the ellipse formula (2) where a and b are the axis halves of the ellipse.

$$x^2/a^2 + y^2/b^2 = 1. \quad (2)$$

To represent a wide variety of crown shapes in transverse section, the equation of Horn(1971) is used and adapted. The vertical transverse sections through the top and the base of the crown and the peripheral points are represented by Eq. (3). The equation of these vertical crown curves is cognate to the formula of the quadratic equation of the ellipse but has a varying exponent E:

$$x^E/a^E + y^E/b^E = 1. \quad (3)$$

Tree base (B) and fork (F) and peripheral points (P_n) are connected with straight lines BF and FP_n (Fig. 4.1). In the case of a crown base lower than the fork, the tree base and the crown base (C) are connected (BC) and the lines FP_n are omitted.
In ground plan the position of standing live or dead trees is indicated by, respectively,

an open and solid circle proportional to the tree diameter. Dead trees lying on the forest floor are shown by straight lines. Micro-relief resulting from the uprooting of a tree is represented with the help of shapes similar to those used for tree crown simulation.

4.2.1 Crown Projection

The crown projection is approximated by the compound of four quarters of ellipses through the peripheral points (Fig. 4.2). The length of the axes of one quarter ellipse equals the absolute difference between the x-values and the y-values of two adjacent peripheral points, e.g. P_1-P_2, P_2-P_3, P_3-P_4, P_4-P_1. The centre of a quarter ellipse is determined by the x- and y-value of the extreme peripheral point in y-direction and x-direction, respectively. Except for the case of equal y-values for P_1 and P_3 and equal x-values for P_2 and P_4 the centres of the four quarter ellipses do not coincide (Fig. 4.2). The complex figure of four quarter ellipses with different centres and axes is nevertheless continuous because the tangents of neighbouring ellipse segments are in line in the peripheral points. The approximated crown projection described above can be transposed onto a horizontal "peripheral plane" at peripheral level above the ground.

4.2.2 Crown Projections of Bending Trees

In the case of bending trees, the position of the tree base may be situated outside the crown projection. When peripheral points of such often narrow crowns are digitized

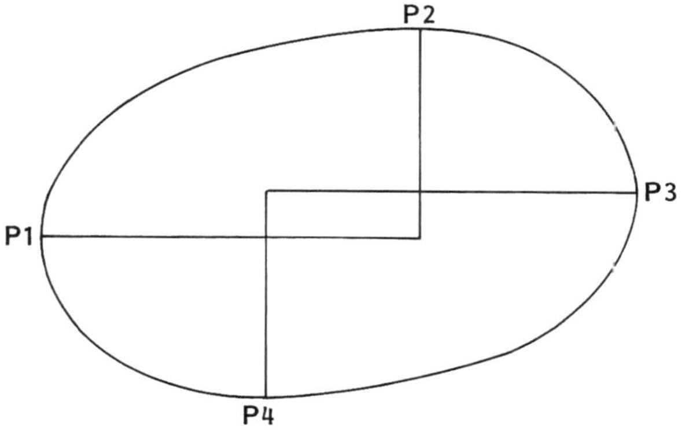

Fig. 4.2. The position of a tree crown in horizontal projection is indicated by a compound of quarter ellipses stretched up by digitized peripheral points, being the extremities of the tree crown in x- and y-direction. The centres of the four quarter ellipses mostly do not coincide

in the above-mentioned way, the model may reproduce an exaggerated crown projection. The degree of exaggeration depends on the angle with the positive x-axis, in which the narrow crown bends away from the stem base (Fig. 4.3). To prevent distortion, these crowns are to be digitized and processed in a rotated co-ordinate system. The new x-axis is stretched between the stem base and the most remote point of the crown, in relation to the stem base. This point at the same time is a peripheral point (P_3). Other peripheral points are digitized in the rotated system as described above. The rotation angle can be derived from the direction of the line from the tree base to the most remote peripheral point (P_3).

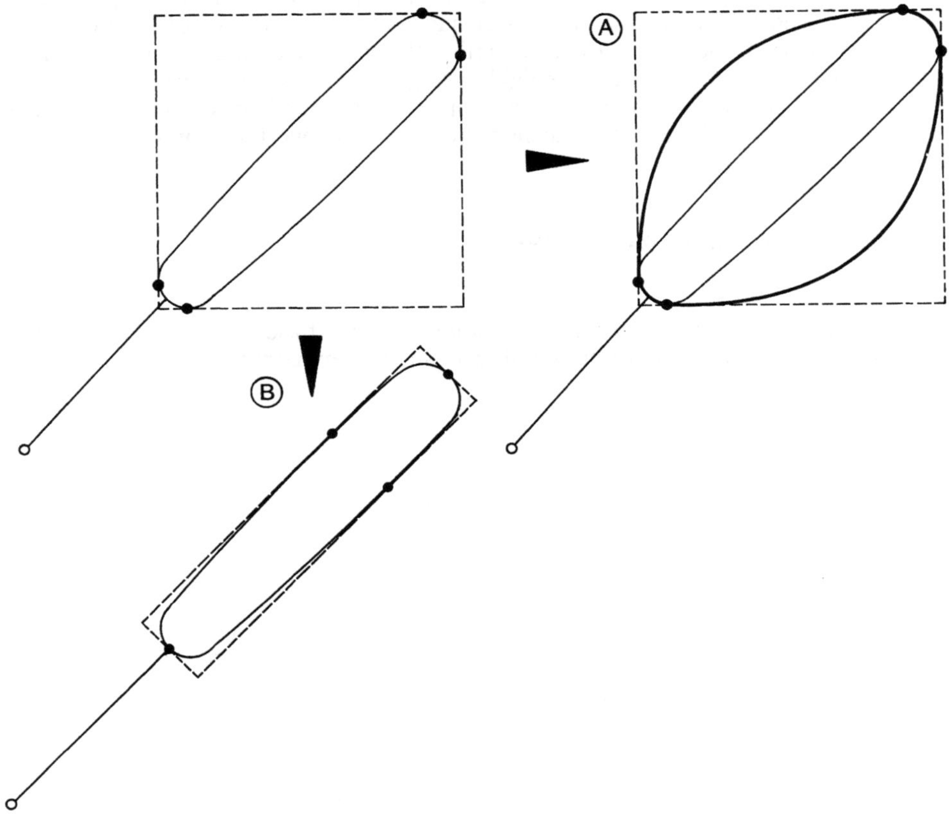

Fig. 4.3. Narrowly stretched crowns of bent trees may become exaggerated, when digitized in the normal co-ordinate system (**A**). Digitized in a rotated coordinate system, with its x-axis parallel to the direction in which the narrow crown bends away from the stem foot, distortion is prevented (**B**)

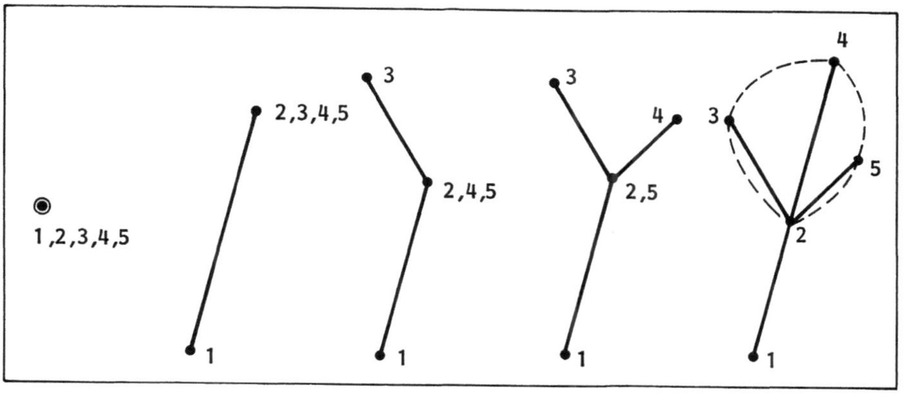

Fig. 4.4. Ends, bends, and bifurcations of fallen dead trees are digitized in the sequence *1, 2, 3, 4, 5*. Their position in the model is indicated as a *straight* line or a *bifurcated* line. Bifurcations of a higher order are omitted

4.2.3 Dead Trees

Standing dead trees in the plan are represented by a solid circle proportional to their diameter. Fallen dead trees if not forked are digitized as a straight line. Forked fallen dead trees are digitized as shown in Figure 4.4.

4.2.4 Calculation of Unknown Co-Ordinates

The peripheral points and the tree base position have measured x and y co-ordinates. Except for the case of the profile drawing in the woody transect area, neither x- nor y-values of the fork and the top and base of the crown are measured in the field. These values are calculated from the eccentricity of the tree base in the crown projection. Eccentricity of the tree base in x-direction (eB_x) can be defined as the ratio of distance between tree base (B_x) and centre point of the crown projection (M_x) and crown radius (Cr_x) as in Eqs. (4), (5) and (6).

$$Cr_x = (P3X - P1X) / 2 \tag{4}$$

$$M_x = P1X + Cr_x \tag{5}$$

$$eB_x = (M_x - B_x) / Cr_x. \tag{6}$$

The values of the eccentricity-index eB_x as a function of the position of the tree base in the crown projection are depicted in Figure 4.5.

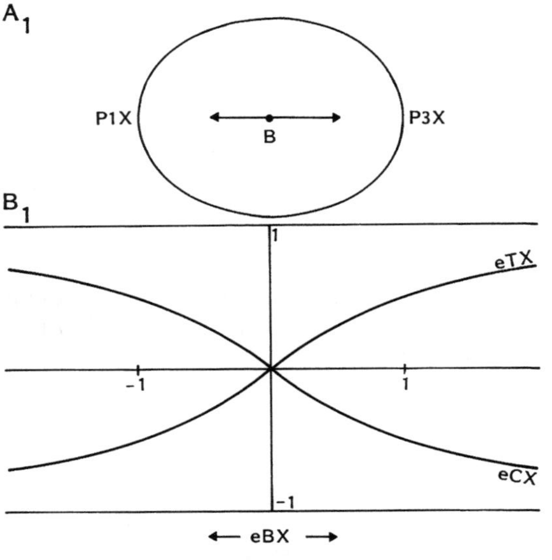

Fig. 4.5. A Position of the tree base (*B*) in the crown projection. The eccentricity index of the tree base $eB_x=0$ if B is situated in the centre of the crown (measured in x direction only). $eB_x=1$ if B is situated on the left crown periphery ($B_x=P1_x$). $eB_x=-1$ if B is situated on the right crown periphery ($B_x=P3_x$). If B is situated outside the crown projection, than $eB_x>1$ or $eB_x<-1$. **B** Graph of the eccentricity index of the crown base (eC_x) and of the top of the crown (eT_x) expressed as the arctangent of the eccentricity index of the tree base (eB_x) [Eqs. (7) and (8)].

The further the tree base becomes situated outside the crown projection, the further T_x and C_x will move towards the crown periphery, each of them in an opposite direction. For trees with great crown depth the shift is reduced. For y-co-ordinates an analoguous procedure is followed.

Because crown top and crown base can never exceed the crown projection, the eccentricity indexes for the top (eT_x) and the base of the crown (eC_x) should lie between plus and minus one. This can be achieved by expressing eT_x and eC_x as the arctangent (in radians) of the eccentricity of the tree base (eB_x), with a reach between $\pi/2$ and $-\pi/2$, multiplied by $2/\pi$, as in Eqs. (8) and (9). Because the shift of the top of the crown is opposite to the direction of the shift of the tree base, eT_x is negated. A reduction factor for relative stem length (R) reduces eccentricity of top and base of the crown when the ratio stem length (= height of the crown base) to total tree height increases [Eqs (7)]:

$$R = (1 - C_z/T_z) \tag{7}$$

$$eT_x = -2/\pi * \text{arctg}\ (eB_x) * R \tag{8}$$

$$eC_x = 2/\pi * \text{arctg}\ (eB_x) * R. \tag{9}$$

The x-values for top and base of the crown can now be calculated (formula 10 and 11) from:

$$T_x = M_x + Cr_x * eT_x \tag{10}$$

$$C_x = M_x + Cr_x * eC_x = M_x\ Cr_x * eT_x. \tag{11}$$

Eccentricities along the y-axes eB_y, eC_y, eT_y are computed in an analogous way. This implies that the x- and y-coordinates of the top and base of the crown generally do not coincide. X- and y-values of the tree fork are determined by the line connecting the tree base and the tree top at fork height (F_z).

4.2.5 Transverse Crown Sections

The crown surface in transverse section through a peripheral point and the top or base of the crown is approximated by the curve generated by Eq. (3). This equation behaves like the ellipse Eq. (2) for positive x- and y- (in this case z-) values. In order to generate complete transverse sections, the positive quadrant of the ellipse has to be reflected to the other quadrants, which have either negative x- or y-values. Axes and centre of the curve generated by Eq. (3) are determined similarly to the ground plan described for the ellipse. All upper transverse sections have one axis (b) in common, equal to the distance from the top of the crown to the peripheral plane. Similarly, the lower transverse sections have a common axis which equals the distance from the base of the crown to the peripheral plane. The horizontal axes (a) differ in length. For the upper and lower transverse sections these lengths equal the distance from the peripheral points to the projection of respectively the top and the base of the crown on the peripheral plane (Fig. 4.6).

4.2.6 Fitting of Transverse Crown Sections

To achieve a best-fitting wrapping of crowns in transverse section, the exponent E in Eq. (3) can be varied. An E-value of one corresponds to a straight line. $E = 2$ results in an ellipse. If $E > 2$ the crown curve will be more convex, becoming rectangular as the exponent approaches infinity (Fig. 4.7).

Although this is not relevant for most broadleaved temperate trees, an exponent less than 1 produces a concave crown curve. Such a curve can be used for tropical tree ferns, e.g. described by Hallé et al. (1978) and Brünig (1976). Thus, with help of Eq. (3), a wide variety of transverse section curves can be generated.

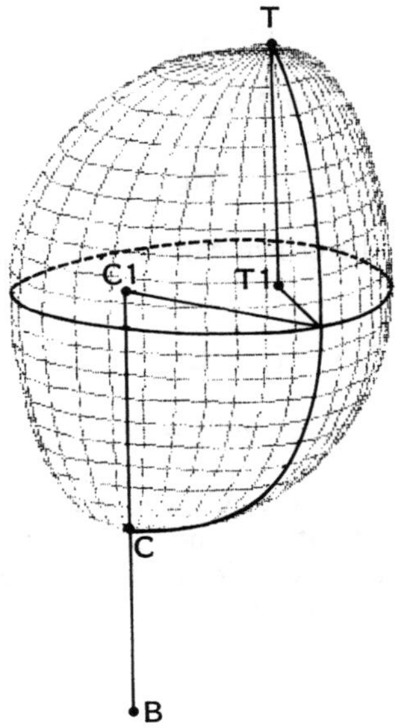

Fig. 4.6. Crown curves through a peripheral point and the crown top have a common axis $T\text{-}T_1$. Crown curves through a peripheral point and the crown base also have a common axis $C\text{-}C_1$. Because the x- and y-co-ordinates of the crown base and the crown top generally do not coincide both axes $T\text{-}T_1$ and $C\text{-}C_1$ are not in line

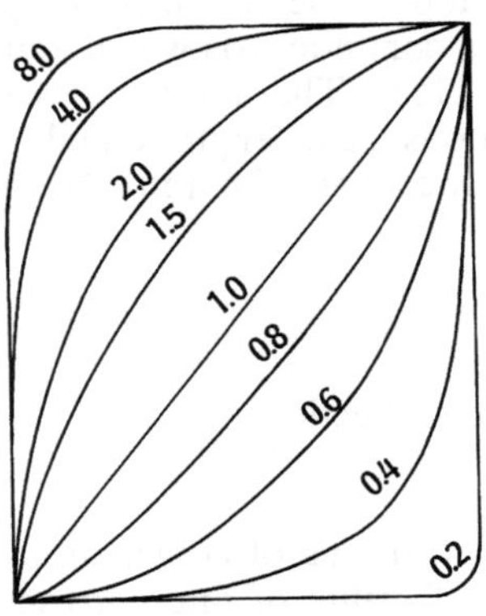

Fig. 4.7. A different exponent E causes the vertical crown curves to become more or less convex or concave

The most appropriate exponent for a fitting wrapping has to be worked out for each individual tree. For this purpose, the crown radius is measured at three-fourths the height between the peripheral plane and the top and base of the crown, respectively. A is the crown radius at three-fourths the height y between the top or base of the crown and the peripheral plane, b the crown radius at peripheral height. The ratio a/b is an indicative value in relation to the exponent of the vertical curves. The solution to the equation $(a/b)^E + (3/4)^E = 1$ [see Eq. (3)] with the unknown E can be an iterative or graphical procedure (Fig. 4.8). However, determination of this ratio for each separate tree in the field would make field work more complex and time-consuming. As a compromise, with help of the hand-drawn transect profile, the exponent is approximated, per species, for the upper crown part (Table 4.1). In spite of high values of variance, because of irregular undulating crown surface and only one measure of a, the E-value for most broadleaved tree species is about two.

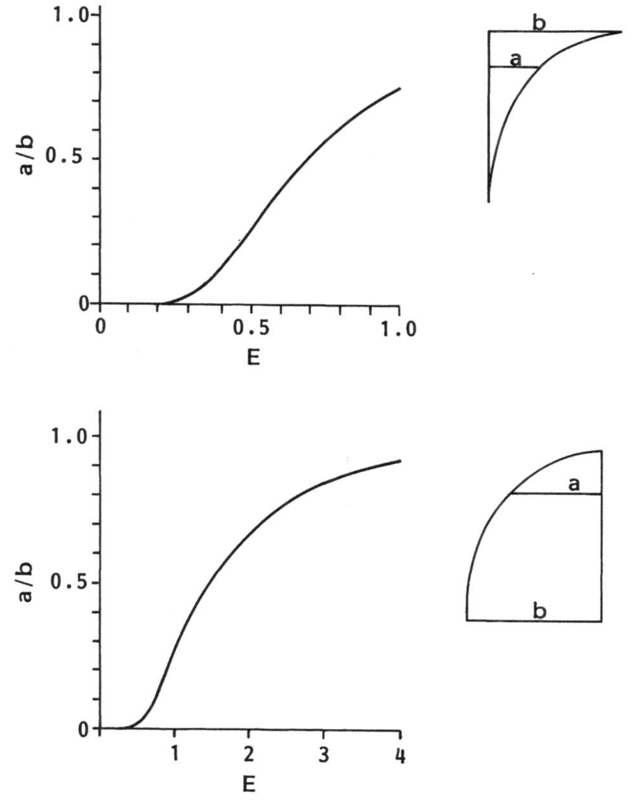

Fig. 4.8. Graph to determine the exponent *E* of a crown curve by measuring the ratio *a/b*. *a* is the crown diameter at respectively three-fourths the depth of the crown curve above periphery height, in the case of spheric crown curves and at one-fourth the depth of a crown curve underneath periphery height in the case of hollow crown curves. *b* is the crown diameter at periphery height. Besides this graphical solution, the Eq. $(a/b)^E + (3/4)^E = 1$ can be solved through an iterative procedure

Table 4.1. Estimated values of mean exponent for the vertical crown curves of different tree species and their variance. For each tree species 100 trees were sampled and the ratio a/b provided the values for the exponent E (see Fig. 4.8).

Species	E	s^2
Pinus sylvestris	1.997	0.446
Fagus sylvatica	2.046	0.304
Quercus robur	2.165	0.205
Fraxinus excelsior	2.229	0.175
Betula pendula	2.256	0.550

4.2.7 Horizontal Crown Sections

Analogous to the vertical curves starting from the four primary peripheral points, from any point on the horizontal ellipses in the peripheral plane, vertical crown curves generated by Eq. (3) can be sent to the crown top or base. All crown curves above the peripheral plane have a common vertical axis in the line connecting the top of the crown and its projection on the peripheral plane. Similarly, all crown curves under the peripheral plane have a common vertical axis in the line connecting the base of the crown and its projection on the peripheral plane. Determination of the co-ordinates of a point on a horizontal crown section at level h above the peripheral plane can be done using the equation of one of the vertical crown curves (Fig. 4.9).

The projected primary peripheral points P_1, P_2, P_3 and P_4 are indicated in the groundplanTherefore the relation between the radius vector x at level h above the periphery plane and the radius vector at the peripheral level is expressed in Eq. (12) by substitution of the height h above peripheral level in Eq. (3) of a vertical crown curve.

$$x^E/a^E + y^E/b^E = 1 \qquad (3)$$

$$x^E/a^E + h^E/b^E = 1. \qquad (12)$$

Formula (12) can also be written as a scale factor in formula (13), expressing the ratio between the radius vector at height h (x) and the radius vector (a) in the peripheral plane.

$$x/a = (1 - h^E/b^E)^{1/E}. \qquad (13)$$

Because all vertical crown curves have a common vertical axis b, the scale factor (13) depends only on height above or below the peripheral plane. Therefore the horizontal crown section at level h above the peripheral plane, that can be described as a set of radius vectors, is the result of uniform scaling of the crown section of four quarter ellipses at the peripheral level. This scaling in the x-y-co-ordinate system is centred in the top of the crown (or the base for the part of the crown under the peripheral plane). Thus at every height level in our model, the horizontal section of the crown is determined by another set of quarter ellipses. Similarly as in the peripheral plane, these ellipses connect the points of intersection of the plane at that level, with the

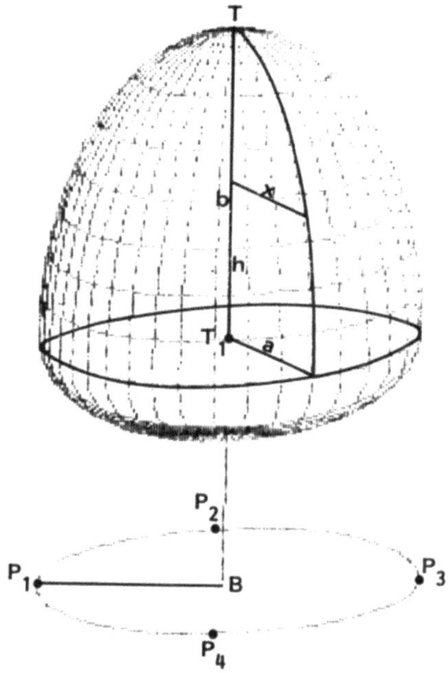

Fig. 4.9. The tree model built with horizontal quarterellipses and vertical crown curves produces a continuous, smooth, well-defined surface, that wraps the volume occupied by the tree crown. The vertical crown curves have an exponent that fits best to each tree species. The common axis *b*, which equals $T\text{-}T_1$, for all vertical crown curves above the peripheral plane is indicated. For one crown curve the radius vectors *x* at height *h* above the peripheral plane and *a* in the peripheral plane are indicated.

vertical crown curves which are stretched, according to Eq. (3), between the four peripheral points and the top or base of the crown.

Consequently all vertical curves generated by Eq. (3) starting from any point at the four quarter ellipses in the peripheral plane, cross all horizontal ellipses at any height above or under the peripheral plane (Fig. 4.9). Contrarily to a surface of revolution, in this model the crown surface is determined by different vertical crown curves that each link up with the compound elliptical crown circumference at the peripheral level. A great variety of physiognomical tree shapes or stylized volumes occupied by trees can thus be described using the same equations for all transverse sections (Fig. 4.10).

4.2.8 Fitting of Crown Projections

The a/b values (Fig. 4.7) measured for a 100 crown projections, regardless species, provided a mean exponent of 2.019 with a variance of 0.405. Generally the planimetered surface of a hand-drawn crown projection differs from the surface that

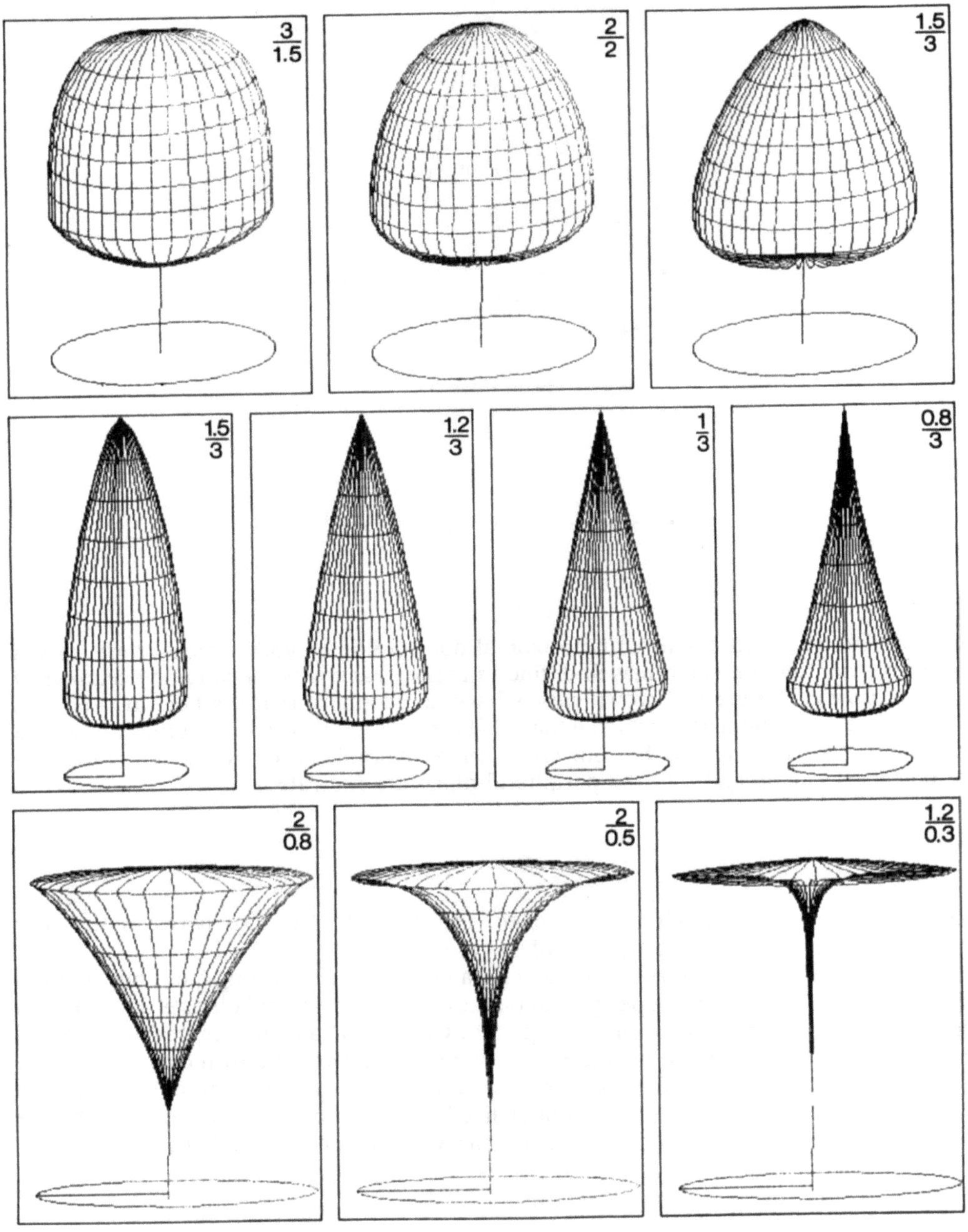

Fig. 4.10. Tree forms generated with different values of the exponent E of the vertical crown curves. The E-value of the upper crown curve (E_U) and lower crown curve (E_L) of the depicted tree shapes is printed in the fraction E_U/E_L

results from the model calculations, because irregularities, particularly indentations in a mainly convex crown circumference, are not accounted for in the elliptical model. The relation between the planimetered and calculated surfaces of crown projections is shown in Figure 4.11 for crowns of different sizes. Projections of angular crowns, frequently occurring in old trees in closed canopies, are exaggerated if the crown angles point in x or y direction. In such cases the original straight sides of the crown circumference between two peripheral points most often become convex.

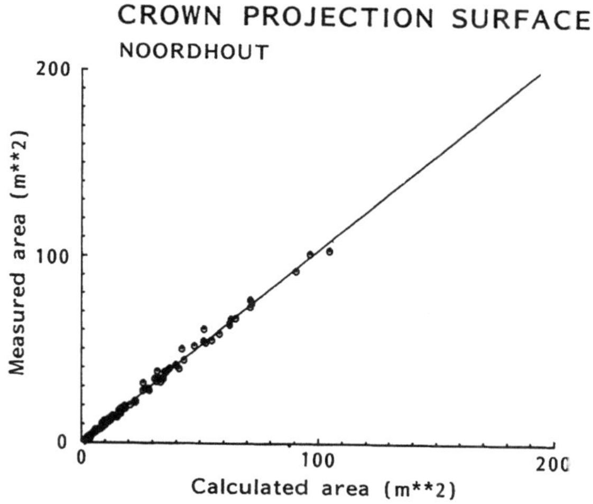

Fig. 4.11. Relation between the surface of the hand-drawn and eliptically computed crown projection in the forest reserve of Noordhout (The Netherlands) with *Pinus sylvestris*, *Quercus robur*, *Betula pendula* and *Fagus sylvatica*. The measured hand-drawn area = 1.029 x calculated area.

4.3 Modelling Uprooting Mound-and-Pit Micro-Relief

Uprootings are drawn in ground plan in the woody transect area and the core area. By making use of the elliptical character of their circumference (Brewer and Merrit 1978; Beatty 1981), uprootings can be represented using a procedure similar to the one discussed for crown projections of bending trees (Chap. 5.4). The mound peripheral points in x and y direction MP_1 to MP_3 are digitized in a rotated co-ordinate system, the x-axis of which coincides with the midline of the trunk (Fig. 4.12). The tree base

(TB) of the fallen tree thus coincides with MP_4. The adjacent pit can be considered as a compound of two quarter ellipses. Two peripheral pit points coincide with the peripheral mound points MP_1 and MP_3. The third peripheral pit point, PP, lies on the outer edge of the pit in the x-direction. The mound as well as the pit are depicted in profile as a compound of two quarter ellipses, by using the height of the mound and depth of the pit and the peripheral points of the ground plan. The spatial shape of mound and pit are simulated just as the upper and lower part of the tree crowns. Because the rounded form of the mound in transverse section is best approximated by an ellipse, the exponent of the vertical "crown curves" for the mound is kept quadratic.

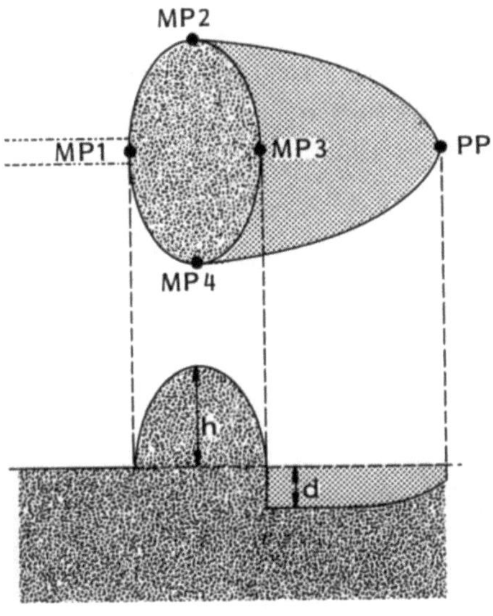

Fig. 4.12. The position of uprooting is indicated as a compound of quarter ellipses. The four extremities in x- and y-direction of the mound in horizontal projection (MP_n) are digitized in a rotated co-ordinate system, with the x-axis parallel to the midline of the trunk, similar to a bent tree (Fig. 4.3). In the same rotated system, the pit is considered as a compound of two quarter-ellipses stretched up between the two mound peripheral points MP_1 and MP_3 and a third pit peripheral point (PP). h is the height of the mound and d is the depth of the pit

Chapter 5

Processing of Primary Tree Characteristics

From the ORACLE database any type of data or combination of data types can be selected. By using the program ORACLEX (see Appendix) a complete standard format file for one object can be retrieved from the database that fits all computer programs developed for data processing. The programs are written in VAX-FORTRAN for data processing and plotting on the computer (VAX and micro-VAX) and in BASIC for the Personal Computer (Olivetti).

5.1 Computing of Crown Area, Crown Cover, Crown Volume and Basal Area

Crown area, crown cover, crown volume and basal area can be computed with the program ARBOSUM (see Appendix). These values are also computed separately for species and for 10 x 10 m blocks of the core area (Table 5.1).

Crown area is calculated as the cumulative area of all crown projections. The crown cover is the cumulative area of all tree crown projections multiplied with their inner crown cover (see Chap. 3). The crown volume is the cumulative volume calculated from the tree crown model. The basal area is calculated from the diameter at breast height.

5.2 Statistical Analysis of Tree Data

The program SILVISTAT (see Appendix) produces computer-plotted distribution diagrams, point plots and regressions. The program includes GENSTAT procedures (Alvey et al. 1982) for statistical processing. For example, regressions can be made of height of fork, crown base and periphery against stem diameter (Fig. 5.1).

5.3 Two Dimensional Computer Plotting of Forest Structure

For the purpose of depiction and analysis of data on forest structure the plot program ARBOPLOT has been developed. Plots of ground plan and profile of the woody transect area as well as of the core area can be computer-plotted with the plot program ARBOPLOT (see Appendix). Plans and profiles in the original core area x-y-co-ordinate system can be directly plotted. For transects at an angle with the core area x-axis, first new co-ordinates in a rotated x-y-co-ordinate system are calculated with the program

Table 5.1. Example of the output of the program ARBOSUM for the core area of the Otterskooi (The Netherlands). Crown area (CA), crown cover (COV), crown volume (Vcr) and basal area (G) are calculated per hectare. The same values are tabulated for the core area blocks of 10 x 10 m and for separate species (codes 14 = *Fraxinus excelsior*, 18 = *Alnus glutinosa*, 25 = *Sorbus aucuparia*). The first two columns under BLOCK indicate row and column of the block. For further explanation see text

Otterkooi (ARBOSUM.SYN)

Synoptic table for volume and area computations

****

Results per ha

Crown area (m^2)	13,495.17
Crown cover (m^2)	8018.61
Crown volume (m^3)	56,276.07
Basal area (cm^2)	233,076.31

****

Results per species per ha

Species	CA	COV	Vcr	G
14	4612.26	2894.38	21,034.68	59,638.03
18	3710.28	1908.63	14,275.62	107,479.86
25	1118.18	835.50	3946.44	13,454.90

****

Results per 10 x 10 m block

Block	CA	COV	Vcr	G
1-1	25.66	15.23	108.38	277.11
1-2	118.55	78.83	508.57	2027.66

NEWTRAN (see Appendix). The beginning and end of the new transect must be entered in core area co-ordinates. The trees to be plotted can be selected as to tree species, architectural code or any other tree parameter (Figs. 5.2, 5.3 and 5.4). Following on a daily field session the BASIC version running on a portable PC provides a visual control on the computer screen of data collected that day.

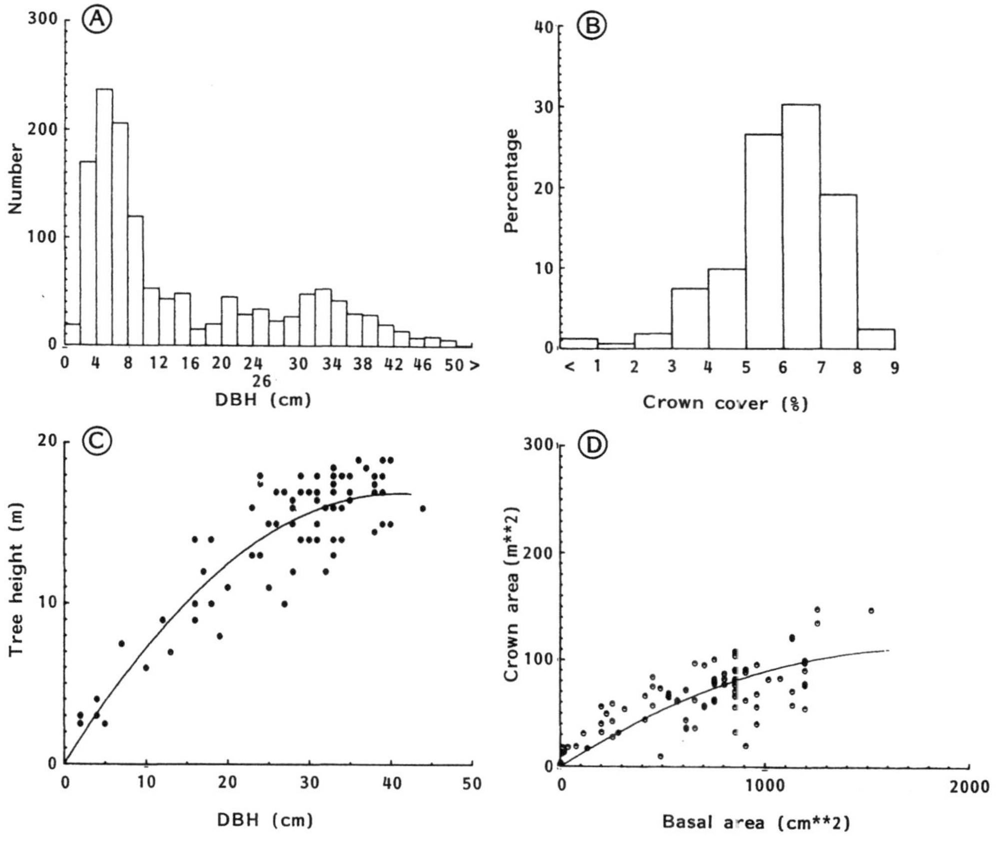

Fig. 5.1. Examples of computer-plotted diagrams and graphics of tree data of the core area. **A** Diameter-frequency diagram. **B** Frequency diagram of inner crown cover. **C** Top height-diameter point plot. **D** Crown area-basal area point plot

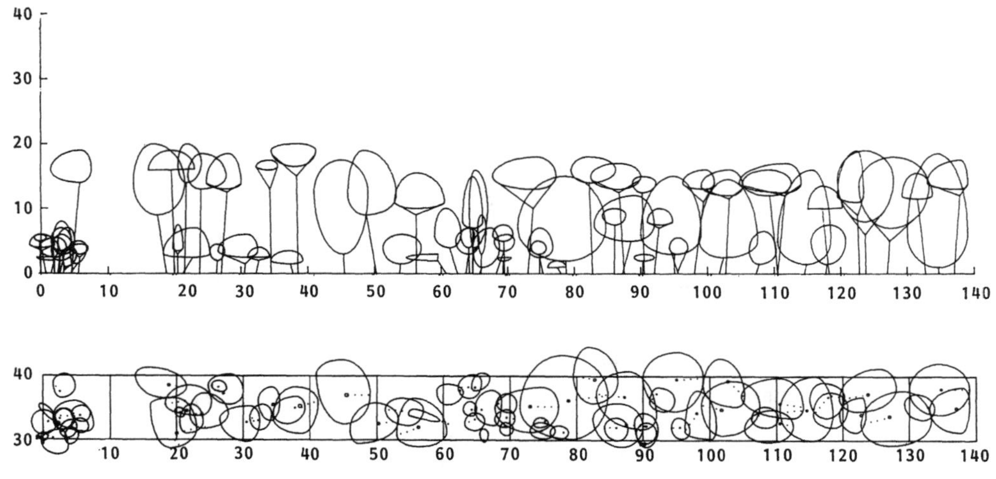

Fig. 5.2. Example of a computer-plotted profile and plan of the transect area of the forest reserve Noordhout (province Utrecht, The Netherlands)

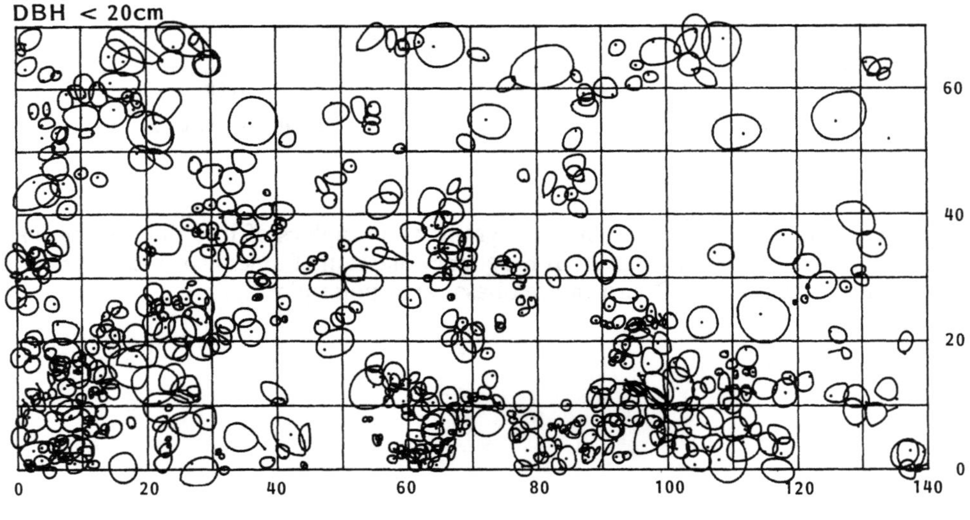

Fig. 5.3. Example of a computer-plotted ground plan drawing of the core area of the forest reserve Noordhout (province Utrecht, The Netherlands). Selective plotting of all trees less than 15 m height shows a characteristic net pattern distribution regeneration under a decay phase of the planted *Pinus sylvestris* stand

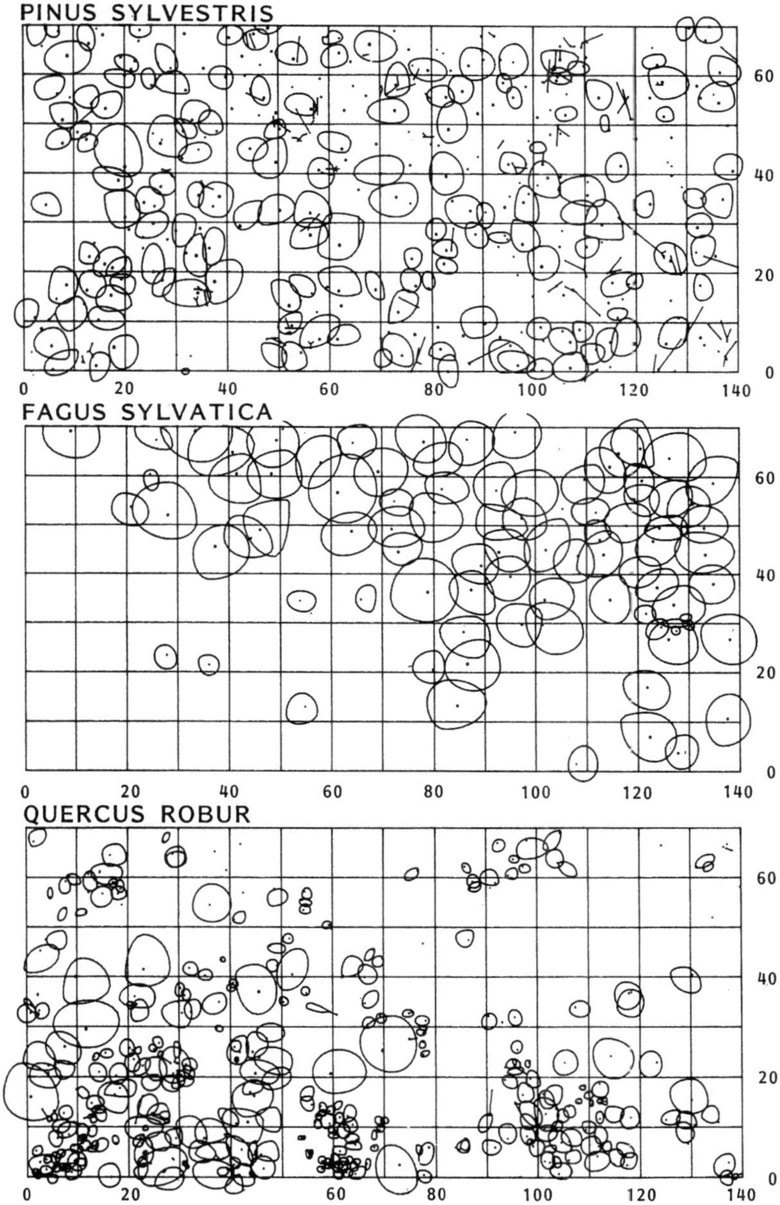

Fig. 5.4. Example of a computer-plotted ground plan drawing of the core area of the forest reserve Noordhout (province Utrecht, The Netherlands). Selective plotting of *Fagus sylvatica* and *Quercus robur* shows that both species exclude each other

Chapter 6

Application Programs

The tree model is not only a tool for two-dimensional depiction but can also be used for a three-dimensional representation of forest structure. In this chapter application programs based on the three-dimensional tree model are presented.

6.1 Light Simulation

Since light is critical to the growth of green plants, the shading by trees in the forest canopy profoundly affects the growth of seedlings, saplings and herbs in the understorey. The light reaching a certain point of the forest floor, or whatever point inside the forest at any other level, depends on canopy cover. Foliage and branches intercept the incident light (Horn 1971; Terborgh 1985).

Evans and Coombe (1959), Anderson (1963, 1964), Bonhomme and Chartier (1971), Bonhomme (1976), Ducrey (1975), Hari et al. (1985) and Oker-Blom (1986) described the use of hemispherical photographs in computing radiation in woodlands. Barkman (1977) used similar techniques with a horizontoscope, used in city building and architecture (Tonne 1954), in studying the microclimate of juniper tree communities (*Juniperus communis*). Proctor (1980), Greis and Kellomäki (1981) and Prince (1987) applied the method to describe bryophyte habitats, to predict tree growth under varying shading and to measure canopy interception, respectively. The other way round, Ondok (1984) uses hemispherical photographs to simulate stand geometry in photosynthetic models. Yoda (1974) describes three-dimensional distribution of light intensity in a tropical forest of West Malaysia. Nakashizuka (1985) describes a model to predict diffuse light conditions in canopy gaps in a beech forest. These authors assume that the shape of the gap is circular and the canopy is horizontal and flat with scattered microgaps among the leaves.

The tree model described above (Chap. 4) can be used to simulate shading patterns determined by the forest architecture. Thus the impact of light availability upon species composition and occurrence and growth of young trees can be modelled. However, the effects of competition for light cannot be separated from those resulting from root competition. It is plausible that when trees die and cause a canopy gap, the root system belonging to those trees also dies. Canopy gaps, especially in zenith direction, also cause an increased throughfall of precipitation. A correlation between canopy cover and occurrence or growth of green plants in the undergrowth can therefore not be explained by light throughfall alone.

The three-dimensional forest model allows computation of the hemispherical cover for fixed points, inside the forest and above soil level, as an indication of the interception of diffuse light and changing oblique radiation. Cumulative cover values can be computed for the separate angles of the hemisphere from all crowns of trees within a certain radius. Light reflected by leaves, stems or branches is not taken into account. For this purpose the program FOREYE was developed (Mansour 1986; Bijlsma 1989). The resulting values of such computation are processed with the subprogram FORFLUX in a similar way as a fish-eye lens- or hemispherical photograph (Anderson 1964; Ducrey 1975; Bonhomme 1976). The fish-eye computations for radiation assessment are made for the 50 adjacent relevés of the herbaceous transect area or in a grid in the ground plan or profile to derive an impression of horizontal and vertical light gradients as determined by the forest structure.

6.1.1 The Program FOREYE

The program FOREYE computes hemispherical crown cover for fixed points at any place in the forest model, over a matrix of 360 azimuthal degrees by 90 zenithal degrees. As stated before, for computation of hemispherical cover, data on forest structure must be available within a certain radius. Diffuse light distribution over a standard overcast hemisphere has been described in an empirical model by Moon and Spencer (1942). According to their cosine law, most radiation can be expected from the angles closest to zenith. Moreover, the oblique radiation coming from the lowest angles above the horizon is intercepted for the greater part on its long path through many tree crowns. The upper angles around zenith are therefore the most important ones for an indication of relative interception of radiation. To what extent the lower angles contribute depends on the actual forest structure. In uniform woods, Horn (1971) limited measurements to canopy coverage in a cone, directly above the view point subtending 10 solid degrees. If larger canopy gaps occur, canopy coverage should be regarded in a wider cone. Oblique radiation may then pass underneath a closed canopy directly above. A cone of 90° around zenith, for a point at ground level, can only be computed if data on trees in a radius of at least one maximum tree height are available. For the core area measuring 140 x 70 m with a maximum tree height of 20 m this means that such computations can only be made for points in an area of 30 x 100 m in the centre of the block. To compute the hemispherical cover, all trees in a 180° cone, within the radius equal to one length of the highest tree in the plot, are scanned for each horizontal and corresponding vertical degree. All individual trees are scanned sequentially for their main stem and crown including their branches.

For scanning purposes, the part of the main stem under the tree fork or under the crown base, in case the fork is situated above the crown base, is assumed to be cylindric. The part of the main stem that starts branching and is situated higher than the tree fork or in the crown is assumed to be a cone ending in the tree top. Per tree the azimuthal left and right limit angles and the vertical angles of horizontal slices of the stems at 0.5 m intervals are calculated (Fig. 6.1).

For the crowns, first the horizontal left and right limit angles are determined, by computing the vertical tangent planes to the crown model. This can be done by computing the tangents and points of contact to the quarter ellipses of the crown projection in the x-y plane, facing to the point of observation or view point (V).

Next, between these limit angles subsequently all horizontal degrees are scanned vertically. All one-degree crown sections to be scanned can be seen as vertical planes, rotating over the previously delimited horizontal angle L_1-V-L_2 (Fig. 6.1a). The minimum and maximum angle covered by the crown is computed through iterative comparison of the angle from V to sequential points at the crown surface.

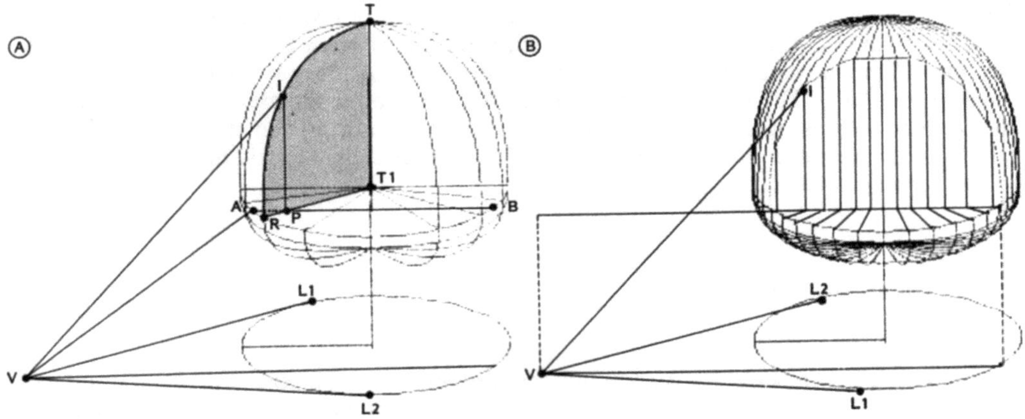

Fig. 6.1. A Computation of a point at the crown surface, as described in the text. **B** Plotting of the vertical section of the rotating plane through the eye point *V* with the tree crown, computed in the way as depicted in **A**

The program takes the following steps to compute these angles.

- A previously defined vertical plane shows a line of intersection with the periphery plane (AB) and two points of intersection with one or two quarter ellipses.
- The point of intersection nearest to V is called A. Because A lies on the peripheral ellipse it is also situated on the outer crown surface.
- The vertical angle made by the line VA is the starting angle for scanning the crown section upwards and downwards.
- To compute the next point on the crown surface along the line of intersection (AB), points (P) are arranged at equal distances. A vector radius (T'P), passing through the projected top at the peripheral plane and a point P, intersects a periphery ellipse in point R.
- The crown surface height is determined by the intersection (I) between the perpendicular line in P and the vertical crown curve RT.
- The vertical angle V-I_n is computed and compared with the angle V-I_{n-1}, the last computed point at the crown surface.

This procedure is repeated until a maximum angle for the upper part of the crown and a minimum angle for the lower part of the crown is found (Fig. 6.1b). Then the next horizontal degree is scanned. When all horizontal degrees per tree are scanned, all values of combinations of horizontal and vertical degrees of the crown circumference seen from point V are stored together with the inner crown cover of that tree. If one of the following tree crowns or stems that is to be scanned covers the same angles, cumulative inner crown cover values for those angles are computed using the formula (14).

$$\text{Cover } (1+2) = \text{cover } (1) + (100\% - \text{cover } (1)) * \text{cover}(2) \qquad (14)$$

The crown shapes are regarded as homogeneously transparent bodies. Because of the visual assessment of the inner crown cover, according to the decimal scale of Londo (see Appendix), neither the path length for the light to pass through the crown, nor the fact that the inner crown cover might change with the angle at which the light passes the crown have as yet been included in the model. The inner cover of the stems is 100 %. In case the point V, for which the hemispherical cover is computed, is situated under a tree

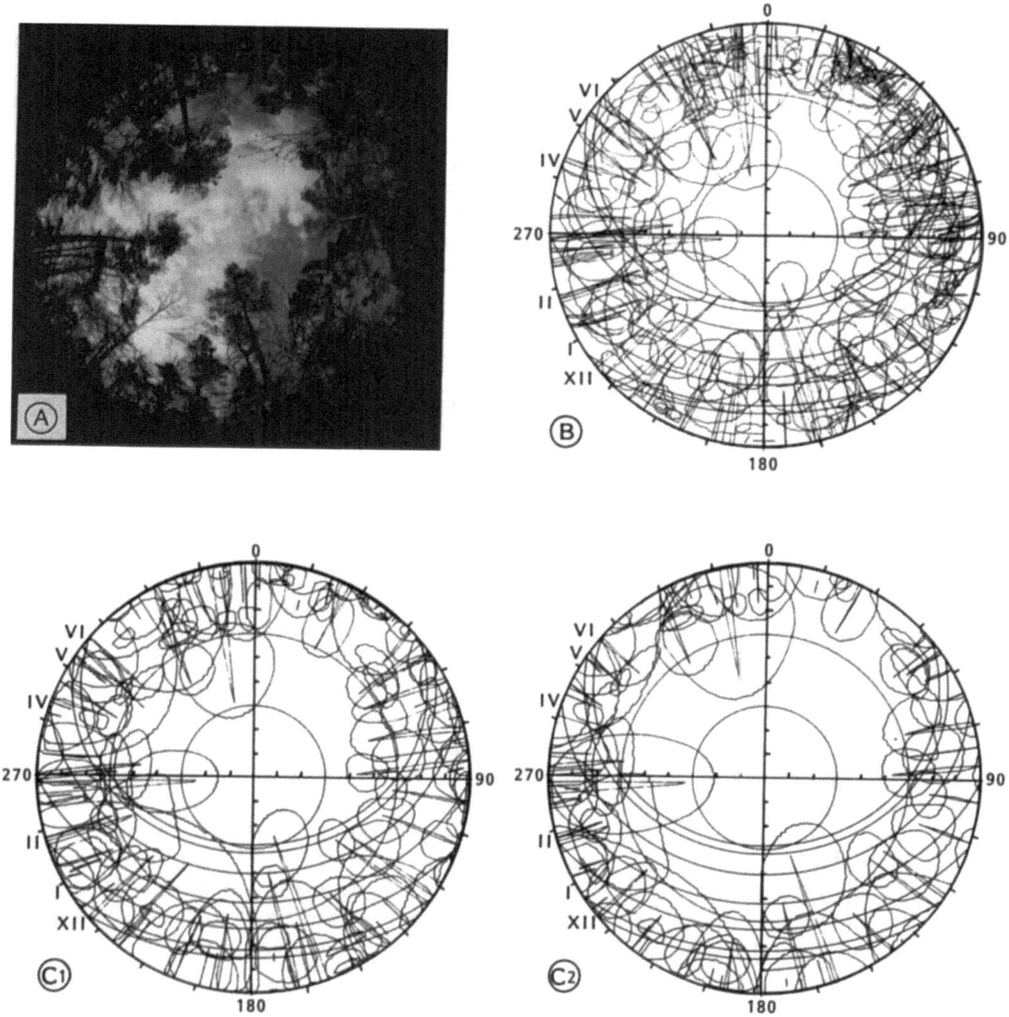

Fig. 6.2. A Example of a hemispherical photograph made in the core area of the forest object Noordhout during winter (broadleaved trees have no foliage). **B** Computer plot of the hemispherical coverage computed for the same core area co-ordinates from the forest model. **C** As **B** but computed for respectively a point 5 m (**C1**) and 10 m (**C2**) above the forest floor

crown (inside its crown projection), only the part below the peripheral level has to be scanned in a slightly different way. The described procedure has also been adapted for the case that the point of view V is not at ground level. In that case, dependent on the relative position of V, below, above or on the crown, and below or above the crown periphery, the tree crown is scanned in a different way. Trees lower than V are skipped. If V is situated between the peripheral level and the top of the tree, only the part above the peripheral level is scanned (Fig. 6.2).

The amounts of direct and diffuse light are computed for a sequence of days by the program FORFLUX (see Appendix). The underlying radiation model is based on standard functions of direct and diffuse light under clear sky and standard overcast sky in J./m/s against sun height (de Wit et al. 1978). The distribution of diffuse light over brightness zones of the hemisphere is determined with the formulas of Ross (1981), distinguishing clear and standard overcast sky.

For the transects in the Netherlands monthly data on hourly values of sunshine duration distributive in percentage at 14 metereological stations over the period 1956-1980 by the Royal Dutch Metereological Institute (K.N.M.I. 1986) are used. Dependent on the hourly values the chance of sunshine at any moment is determined. There is an option for random or fractionary determination of hourly sunshine duration. The amounts of direct and diffuse light can be computed over any specified period of the year. However, because the data on inner crown cover concern summer assessments in the period May-September, only computations for this period are reliable.

6.2 Vegetation Data Processing and Forest Structure

For processing of vegetation data and environmental data, standard software packages are available, e.g. TWINSPAN (Hill 1979b; Jongman et al. 1987), DECORANA (Hill 1979a and Hill and Gauch 1980; Jongman et al. 1987), ELLENBERG (Ter Braak and Gremmen 1987)and CANOCO (Ter Braak 1986, 1987; Jongman et al. 1987).

Data sets of sequential relevés of the same herbaceous transect area are combined for processing with the programs TWINSPAN and DECORANA. Relevés made in different years of one quadrat can be assigned by TWINSPAN to different clusters. Similar cluster transition sequences for quadrats indicate similar vegetation changes in those parts of the transect area. The scores of the relevés on the first two DECORANA axes are plotted. A delineation for relevés belonging to the same TWINSPAN cluster is drawn. Transition vectors that connect different year relevés of the same quadrat are drawn to indicate shifts of quadrats through the diagram. Similarity in vegetation changes in different quadrats can be deduced from the direction of the transition vectors.

To relate vegetation changes to the forest structure of the core area, FOREYE computations of the amounts of direct and diffuse light are made for the 50 adjacent relevés of the herbaceous transect area. The amounts of direct and diffuse light are used as an environmental factor in the canonical correspondence analysis (Ter Braak 1986, 1987).

6.3 Aerial View Simulation

Mapping of tree species in the site area can be done through processing of digitized aerial photographs. The forest model can be used as a ground reference for identification of tree species on remote sensing images. An aerial view can be reconstructed from the data collected in the core area or the woody transect area. The height of the upper crown surface is computed for each pixel in a grid. Pixel size of the reconstructed aerial view can be transformed to that of a digital remote sensing image that covers the plot. For all tree crowns that cover the same pixel the heights of the crown surfaces are compared. The height of the highest crown surface in the pixel and its additional tree characteristics are stored.

The result of this procedure is an orthographic parallel projected grid of highest tree crowns. Because this computed orthographic projection has to be collated with the aerial photograph, which is a central projection, transformation of the first to a central projection is necessary. Thus, each pixel can be assigned to the data collected on the individual tree that was measured and classified in the field. The corresponding pixels of the remote sensing image and the transformed aerial view serve as a large training set of pixels for interpretation of all pixels beyond the plot (Schowengerdt 1983).

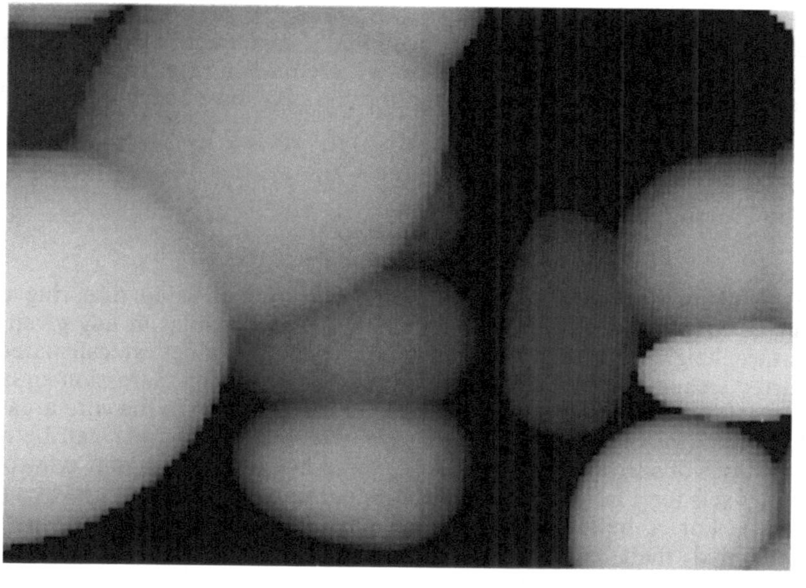

Fig. 6.3. Upper canopy height computed with the CANOPIX program. Pixels of increasing canopy height are depicted in increasing grey-value.

6.3.1 The Program CANOPIX

A procedure to compute the crown surface height at a given point, similar to the one used in the program FOREYE, is followed in the program CANOPIX (Bijlsma 1989). Because of computation of the canopy height in a rectangular grid, forest structure must be scanned in parallel vertical planes at equal distances. The further procedure is identical to the scanning procedure in the program FOREYE. Again trees are scanned individually. In each vertical plane at the line of intersection (AB) with the periphery plane the points (P) are arranged at equal distances. A vector radius T'P again intersects a periphery ellipse in point R. The crown surface height is determined by the point of intersection (I) between the perpendicular line PI and the vertical crown curve RT (Fig. 6.1a). Contrarily to FOREYE, the crown surface heights of all trees in each one pixel are compared. The highest crown surface per pixel is stored. Finally, for every pixel the tree number, its crown surface height, the tree species, the inner crown cover and vitality indices of the tree concerned are stored. Figure 6.3 shows the result of computation of the upper canopy height. Pixel height is shown as an increasing grey-value.

6.4 Forest Reconstruction with Tree Ring Data

Tree ring analysis can be used for reconstruction of individual tree growth in uneven-aged forests in climates with a growth-arresting season. Analysis of diameter growth of uneven-aged trees in an irregular spatial arangement results in a complex data set that can hardly be put in such an order as to communicate it to readers. The tree model developed in this study can be used for visualization of such complex data sets.

6.4.1 The Program GROWRING

Tree diameters for any year can be reconstructed using tree ring cores. For depicting the development of an individual tree, its dimensions at any given time, e.g. crown surface, height of the top, periphery and crown base values are calculated from the reconstructed tree diameter increment. This can be done using regression equations that link the present tree dimensions to tree diameter, as measured in the core area (Fig. 6.4).

Uprooting or die-off of present standing or lying dead trees can be dated with growth reactions of surrounding trees. Age determination of trees that grow on mounds of uprootings sets for a minimum time period since uprooting (Oliver and Stephens 1977; Zeide 1981). For a better impression of past forest structure, dead trees can be "re-erected" and their dimensions in further reconstruction can be worked out proportionately to trees of about the same size in a comparable position.

Calculated, reconstructed dimensions of all trees in the sample plot are for each year written to year-files. A sequence of plotted year-files visualizes, from the diameter increment, the growth of the trees and enables interpretation of competition between trees due to their positions and dimensions.

This way of depicting tree and stand development in the past still has some restrictions and therefore the images that are created should as yet be interpreted with caution:

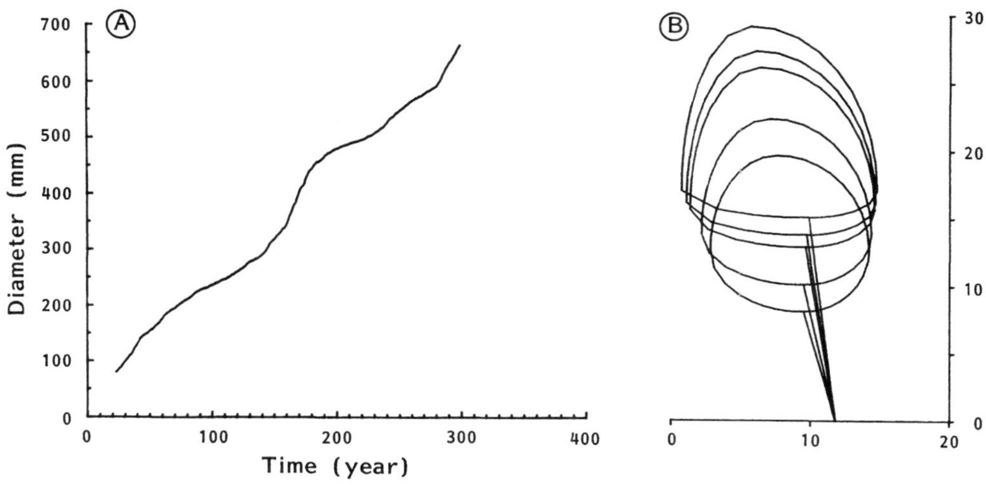

Fig. 6.4. A Diameter growth of a beech (*Fagus sylvatica*) in Fontainebleau. **B** Reconstructed tree dimensions at sequential time intervals for the same beech. Height of the crown top, periphery and base and the crown projection surface have been computed with regression equations

- Regressions of present tree dimensions with tree diameter are based on measurements of trees thicker than 5 cm at breast height. These regressions can be used to determine dimensions of trees with reconstructed diameters of more than 5 cm breast height. In the case of reconstructed diameters less than 5 cm, tree dimensions should be computed deviantly. Here a linear decrease proportional to diameter, from the regression value for 5 cm Dbh on to the origin, is used for all tree dimensions.
- The height at which the samples of tree rings are taken restricts the phase of development of a tree that can be traced back. Because trees were sampled at breast height, they appear in the retrospective view at the moment they surmount the height of 1.30 m and already have a certain crown dimension.
- Crushing of part of the tree crown by the fall of another tree, or a gradual die off and shrinking of a tree crown is not included so far. For a reconstruction of a tree crown, before the moment it became damaged or its dimensions started to decline because of senescence, its original crown dimensions have to be deduced from regressions. Therefore, in the model, senescent or damaged trees that lost part of their original crown will be assigned to an initial crown size according to the diameter regressions. The moment of starting decline can be retraced from the tree rings and can sometimes be associated with extreme weather conditions.

Only trees for which the tree rings could be sampled can be traced back and depicted in the retrospective view of forest structure.
- Trees of which no trace is left (died and completely decayed) may at some time in the past have dominated the forest canopy. This is realistic because trees moulder away after some decades (Koop 1981). In flooded forests, recently fallen trees may have drifted away. On these sites even the relatively "hard memory" of an uprooting mound and pit, that normally remains visible for centuries, is leveled and washed away within a few years.

- Rings in the heart of the tree can often not be measured. Trees may rot at the core and become hollow, the wood at the heart may be too soft to be sampled with the borer, or the year rings have become unreadable. Alternatively, tree rings in the heart of the tree may not have been measured because the heart was missed in the sampling. In such cases data on the inner tree rings, that give information about the absolute age and youth growth of the tree, are missing. In our reconstructed, retrospective view these trees "disappear" for further reconstruction purposes beyond the year of their last measurable tree ring.

Part C Examples of System Application

In part C of this study examples are given of the applications and use of the information system. Although time series are not yet long enough to show long-term and thereby large shifts in forest mosaics or species composition, the examples given may demonstrate the potential of the methods for such applications as well. The examples are presented in the following chapters, in a sequence from (1) retrospective study, an attempt to reconstruct long-term forest dynamics, first cyclic and second successional dynamics (Chaps. 7 and 8); (2) monitoring study, using time series of sequential forest inventories (Chap. 9); (3) comparative study, analyzing variables across many forest objects (Chap. 10) and; (4) simulation study, simulating impacts of trees upon each other and upon herbaceous vegetation and tree species regeneration (Chap. 11).

Chapter 7

Reconstruction of Long-Term Cyclic Forest Dynamics

In this chapter, reconstruction of long-term forest dynamics will be demonstrated, using the data collected in some of the forest objects under study that have a relatively long history of spontaneous development with minor human interference. Two forest reserves have been selected as examples of reconstruction of cyclic developments in European lowland forest. A third forest gives an example of deviation of cyclic processes under grazing pressure. Present forest structure and species composition, tree measures and architectural codes collected for the individual trees in the core area (Chap. 3) in combination with historical maps and written documents descriptions of former use (Chap. 3) can explain much of its developmental history. Tree rings can be used to reconstruct forest growth. Dead wood and traces of uprooting (Chaps. 3 and 4) sometimes give an indication of forest dynamics that dates even further back than the tree ring information.

7.1 Cyclic Forest Dynamics in the 600 Years' Unexploited Beech Forest of Fontainebleau (France)

The strict reserve of la Tillaie (34 ha) is part of the 17,000-ha state forest of Fontainebleau. The reserve is situated on a plateau, 135-140 m above sea level, in a hilly country that geologically belongs to the basin of Paris. The climate is of a maritime type with some continental influences. The subsoil of limestone is covered with a sand layer of 0.3 to 2.0 m thick. Dependent on this depth, different soil types have developed, each type carrying its own characteristic vegetation (Lemée 1978).

In the thin sand cover, up to 80 cm, calcareous brown earths occur with a woodmelick-beech forest (*Melico-Fagetum*) according to Trautmann (1972) and Van der Werf (1989). The *Melico-Fagetum* is a pure beech forest (*Fagus sylvatica*) in which *Acer campestre* and *Fraxinus excelsior* occur in small numbers in regeneration groups under canopy gaps. The herbaceous layer is dominated by *Brachypodium sylvaticum*, *Melica uniflora*, *Festuca heterophylla* or *Ruscus aculeatus*. Under canopy gaps *Calamagrostis epigejos*, *Brachypodium pinnatum* and *Rubus fruticosus agg*. occur.

In deeper sandy profiles, leaching of iron and humus led to the development of a podsol. Vegetation on these podsols can be characterized as a sessile oak-beech forest (*Fago-Quercetum*). Besides *Fagus sylvatica*, a small number of *Quercus petraea* can be found. A thin shrub layer is formed by *Ilex aquifolium*. Characteristic species for the thin herbaceous layer are *Deschampsia flexuosa*, *Carex pilulifera* and *Lonicera periclymenum*, while characteristic species of the *Melico-Fagetum* are absent. Clearings are mostly overgrown by a dense vegetation of *Pteridium aquilinum*.

In 1982 two core areas (70 x 140 m) were charted. Core area I represents the *Melico-Fagetum* and core area II the *Fago-Quercetum*. The regeneration mosaic shifts over 15 years could be retraced by comparison with tree-crown mapping (scale 1:1000) by Bouchon et al. (1973) and were published by Van Baren and Hilgen (1984) and Koop and Hilgen (1987). In 1987, when herbaceous transect area relevés were reassessed, tree rings were sampled in the woody transect area, together with a central square area measuring 40 x 40 m. One hundred trees with Dbh > 10 cm were cored twice at breast height. The results were processed as described in Chap. 6.

7.1.1 Historical Analysis

Unwritten Data

Pollen analyses (Guillet and Robin 1972; Lemée 1981) had revealed that since the eighth century the forest area developed from an open grassy oak forest to the present almost pure beech forest. ^{14}C Dating and species identification from charcoal remnants by Jaquiot et al. (1973) confirmed that La Tillaie was an exploited oak forest during the eighth century. The name la Tillaie refers to *Tilia*. At the present only few *Tilia cordata* trees have survived and pollen analysis reveals that since the eighth century *Tilia* has played only a minor role in species composition of the forest.

Clear-Cut in 1372

In the fourteenth century, the forests of Fontainebleau became part of the heritage of the king's widow. Part of the forest was cut as a source of income (Tendron 1983) including the forest compartment named "Coup des Reines" near the present forest reserve. A document of 1372 mentions the sale of high trees in the compartment of la Tillaie, probably for building purposes in Paris (Grand-Mesnil 1982). In the forest of Fontainebleau old broadleaved high forests have been preserved as a royal hunting field and to provide a scene of ancient woodland around the palace of Fontainebleau. As compensation for the damage caused by the high numbers of deer, local inhabitants gained the right of cattle grazing and collection of dead wood and litter (Tendron 1983).

Management Plan of 1664

The first management plan (Barillon d'Amoncourt, 1664) describes all forest compartments and their subdivisions in great detail. The forest sites "Petite Tillas" and "Grande Tillas" described in this plan can be traced on the map of Mathis (1725). The individual subdivisions of both forest sites can be located quite accurately because embracing forest sites, adjacent roads, the size of areas and sometimes soil conditions are mentioned. As can be seen from Figure 7.1, the present forest reserve is covered by the subdivision D and E of "Petite Tillas" and subdivision B of "Grande Tillas". Plot I is covered by subdivision B and plot II by subdivision E. Part of the management plan (Barillon d'Amoncourt 1664) concerning both plots has been translated below.

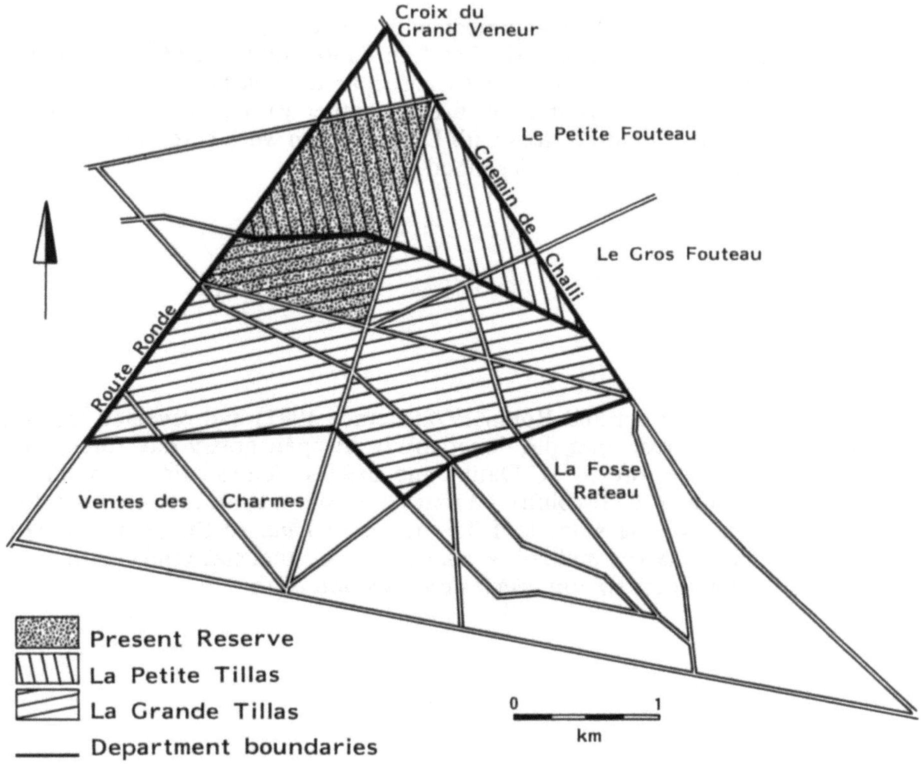

Fig. 7.1. Topography of La Tillaie after the map of Mathis (1725) with allocation of the forest sites "Grande Tillas" and "Petite Tillas" and their subdivisions according to the descriptions of Barillon d'Amoncourt (1664)

"Grande Tillas

Embraced on one side by La Petite Tillas and on the other by La Ventes des Charmes, on one end by La Route Ronde and at the other by La Montagne Pierreuse and Fosse Ratault. Containing 140 arpents (one arpent = 1/2 to 1/3 ha) 46 perches (one perch = 34 m^2) to wit:

B 129 arpents 68 perches of an old high forest of old beeches of 200 to 300 years old, thick and heavy, oaks and some lime trees of the same age and old wide-spread standard oaks of 300 to 350 years old, underneath an abundance of beeches and some lime trees of different age from 35 to 42 year to 200 years. Some parts are rather open because of the fall of old beeches and some wind-thrown oaks. The area adjacent to La Route Ronde is the best and with many beautiful oaks on the best soils. Concerning the second part (subdivision B) that regenerates, although very irregular, it will maintain itself in this way, and one can for forest aesthetical reasons dispense with any cutting for a long time.

Petite Tillas

On one side Le Chemin Chailly, on the other La Grande Tillas, on one end La Fosse Ratault and on the other La Route Ronde. Containing 94 arpents and 80 perches to wit:

D 12 arpents 64 perches along La Route Ronde badly occupied by oak as well as hornbeam of 150 years old, the whole undergrown with plants of hornbeam and beech.
E 62 arpents 47 perches of oak of 100 to 150 years and 200 years old on a good soil although gravelly where the beech is not able to succeed, because it sends its roots between two soil layers and is blown down by the big storms, but underneath well occupied by the same species beech of different age, the youngest succeeding the old, the open places being filled up quickly. It is not necessary for a long time to plan any felling in this part of the forest, that can always maintain itself entirely while remaining well occupied."

Management Plan of 1716

Because of fast forest degeneration, the second management plan (De la Faluere 1716) prescribes artificial rejuvenation of open stands by planting and sowing. La Tillaie is described as a collapsing forest with many dying trees and open spaces. The oldest beeches in la Tillaie, however, date from about 1600, and a continuous tree ring increment of old oaks since the 16th century (Lemée 1966) confirms the existence of La Tillaie as an ancient woodland, far older than the second management plan. With Tendron (1983), we may conclude that the second management plan was never carried out in la Tillaie. Together with some other old stands, la Tillaie was last cut in 1372 and has developed spontaneously ever since.

1853 Establishment of Strict Forest Reserves

Many paintings of the school of Barbizon depict the forest of Fontainebleau in the 19th century as a high forest (Sillevis and Kraan 1986). In 1837 painters and tourists rose against planned cuttings for rejuvenation of the oldest stands of Fontainebleau including la Tillaie (Tendron 1983). Strict forest reserves were established already in 1853 for aesthetic reasons. Ever since, wood harvesting has been prohibited. These forests belong to the oldest well-protected nature reserves in Europe. Lacordaine and Lescuyer (1853) describe the old stands in la Tillaie as follows:

"High forest of beech and oak with some hornbeam 100 to 300 years old, some trees even older, forming a very beautiful massif, very high and still healthy and complete especially in subdivision 171 (the present reserve la Tillaie). One also meets open parts and places where the stand does not seem to be older than 100 years. In the open and badly occupied places, seedlings seldom exist of the same age that could be conserved in the case of a regeneration cut."

More generally, Lacordaine and Lescuyer describe the stands of the reserves from a forester's point of view as:

"The old high forests have unfortunately all reached their period of decline. Even in the most beautiful parts most of the trees show signs of weakening. Generally speaking the greater part of the trees have partly dead crowns and are attacked by a rotting process from the top to the tree base. Most of these high forests are unexploited, because of the almighty veto of the artists and tourists. Each year, only trees that dried on the stump, rotted and fell under their own weight are removed. So they become thinned out day by day. The emptied stands and the already numerous empty places are invaded successively by juniper and bracken. In those exceptional cases where some natural regeneration establishes itself, generally incomplete, mixed up with dead wood of very irregular age and for a long time suppressed, it does not have any future. The natural regeneration that sometimes establishes itself, and which is good and complete enough, can only die for the greater part under a lasting canopy cover and successively all the high forests meet with their decline in the near future."

7.1.2 Tree Age Distribution

Two tree generations can be recognized in the tree ring data of core area I. The oldest generation dates from 1651 to 1693 (Fig. 7.2). Except for one tree from 1785 there is a regeneration gap in the period 1693-1851, after which trees established almost continuously until 1950.

In core area II the same two generations of trees can be recognized. The oldest canopy trees established in the period 1609-1723. Then it takes 146 years before the second generation appears. The latter became established almost continuously with lags less then 15 years since 1869 until 1963 (Fig. 7.2). Trees younger than those of 1950 in core area I or 1963 in core area II are probably thinner than 10 cm and were not sampled for tree rings. No particular spatial pattern in different age of the old-generation trees can be recognized.

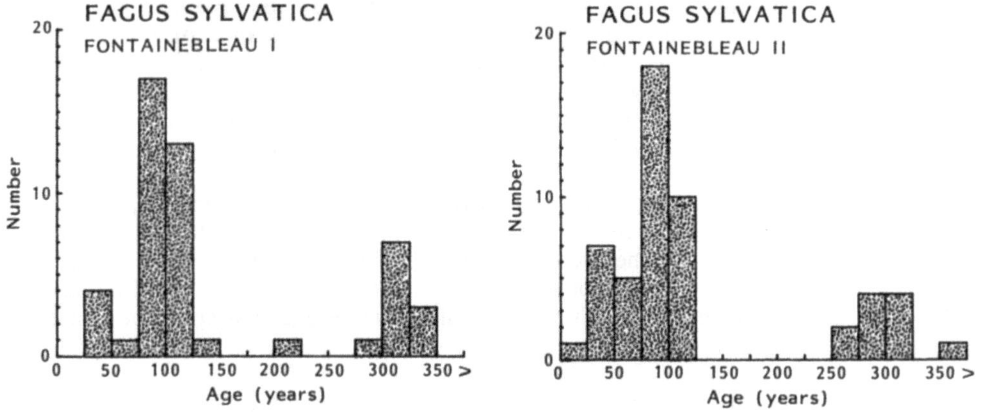

Fig. 7.2. Age distribution diagrams of all trees sampled for tree rings in core area I and II. Tree age is established by counting the tree rings. If the tree core was not in the sample, the remaining tree ring numbers were estimated as the distance to the tree core divided by the average tree ring width

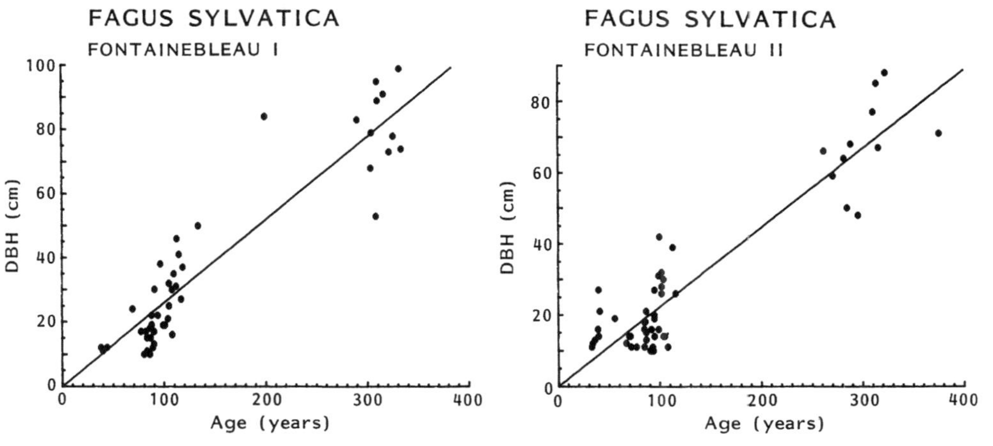

Fig. 7.3. Relation between tree age and Dbh for all trees sampled for tree rings in core area I and II

The generation gap of 150 years recognized in both sample areas rests on tree ring counts of only 100 trees in a restricted area of 0.5 ha. Nevertheless, it can probably be concluded that it holds for the rest of the core areas and the reserve as well. The relation between tree age and diameter is plotted in Figure 7.3. The gap in tree age is reflected in tree diameter distribution of the trees sampled for tree rings. The dip in this tree diameter distribution due to the generation gap is centred around Dbh=50 cm. In the diameter distribution diagrams of the separate core areas, these dips can easily be recognized (Koop and Hilgen 1987), as well as in the diameter distribution diagram of Lemée (1978), representing the whole forest reserve.

7.1.3 Pattern in the Regeneration Mosaic

When looking at the pattern of canopy height of both core areas in Fontainebleau it is striking that the lower trees are spatially separated from the big trees. They are arranged in circles around the big trees of a biostatic patch, showing a network woven between the highest trees (Figs. 7.4, 7.5, 7.6, 7.7).

Regeneration units underneath larger open spaces between the highest trees can be seen as nodules in the network. The narrowly stretched canopy gaps can simply result from sequential fusion of one-tree gaps. However, falling trees, slashing a second canopy gap where the tree crown hits the ground (Oldeman 1974b) and the "bowling" effect of one tree falling against another, can enhance this pattern of narrowly stretched canopy gaps. The remaining old trees expand their crowns in those gaps, but meanwhile trees that were waiting underneath try to reach the canopy. The height-diameter ratio of the ingrowing tree is very high and does not decline until the tree has established itself in the upper canopy. These extremely slender trees primarily invest in height growth in the "chimneys", i.e. narrow funnels or cylinders under a slight gap in the canopy through

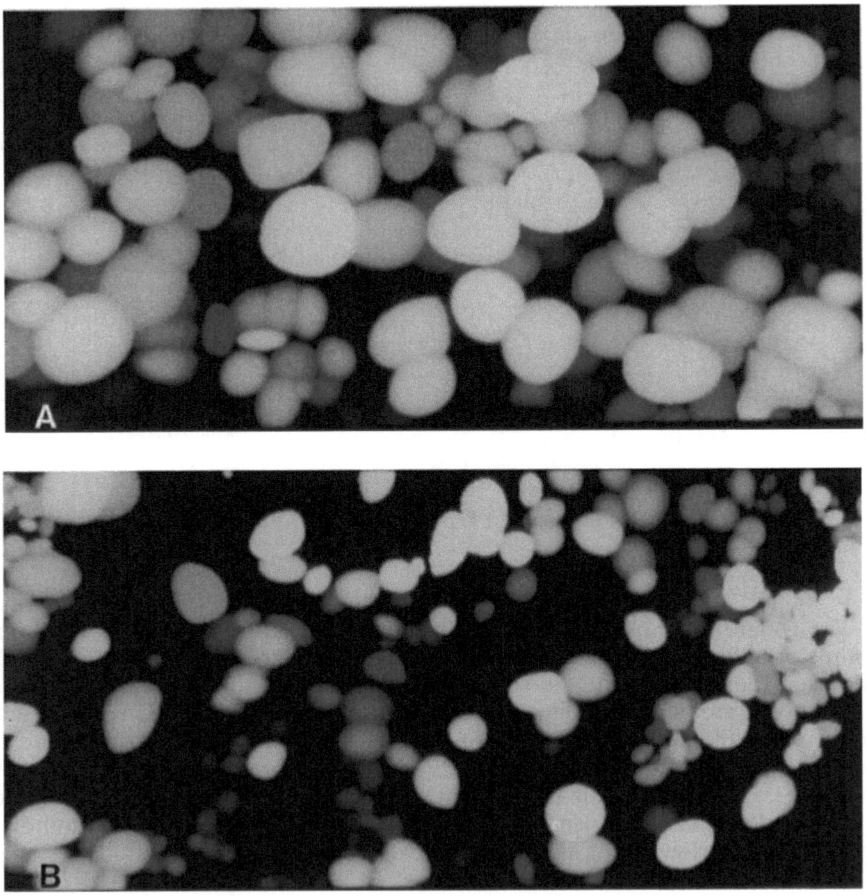

Fig. 7.4. Upper-canopy height computed from the forest model in the core area I in La Tillaie (Fontainebleau). **A** Upper-canopy height of all trees, showing the fragmentating canopy of first generation trees. **B** Upper-canopy height of trees lower than 20 m, clearly showing the pattern of the network of aggrading regeneration units surrounding the emerging first generation trees

which the light energy level is raised (Oldeman 1974a; Hallé et al. 1978). Meantime these trees can remain erect because of their wind-sheltered position underneath the canopy.

Delimitation of the regeneration units as being "a surface unit on which at one moment in time a vegetation development has begun" (Oldeman 1983, 1989) is difficult. The trees of the oldest second-generation units already reached the canopy of the old first-generation trees, although there is a generation gap of 150 years between them. These groups of newcomers can still be recognized by higher height-diameter ratios, narrow crowns and a higher tree density. The limits of younger regeneration units sideward to the old trees are clear. Lengthways a delimitation is less evident because of

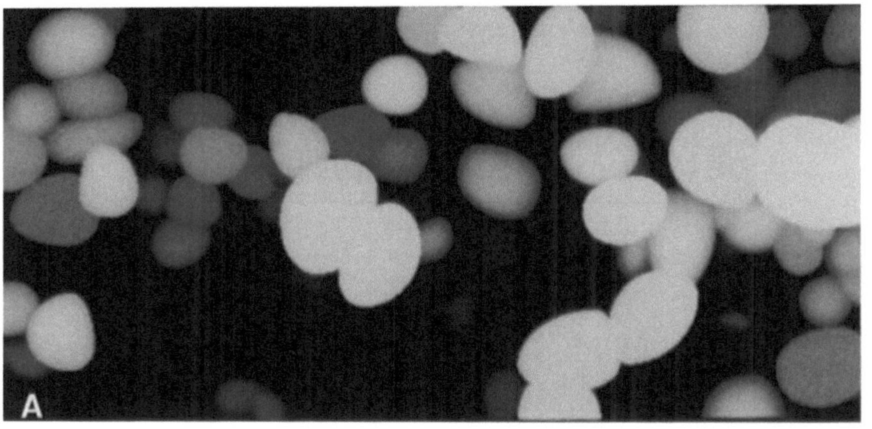

Fig. 7.5. Upper-canopy height computed from the forest model in the core area II in La Tillaie (Fontainebleau). **A** Upper-canopy height of all trees, showing the fragmentating canopy of first generation trees. **B** Upper-canopy height of trees lower than 20 m, clearly showing the pattern of the network of aggrading regeneration units surrounding the emerging first-generation trees

a progressive and more or less continuous fusion of units. Two factors favour the enlargement of regeneration units, which may obscure the network pattern described above.

- As shade tolerance of young trees decreases, for example on nutrient-poorer sites, tree regeneration in narrow gaps is delayed and must wait for gap enlargement.
- Many old trees may lose their vitality at the same time and the decline of an old stand takes place in a rather short time period over larger areas.

Fig. 7.6. Side views of seven adjacent 10 x 140 m transects through the core area I in La Tillaie (Fontainebleau) plotted with the program ARBOPLOT

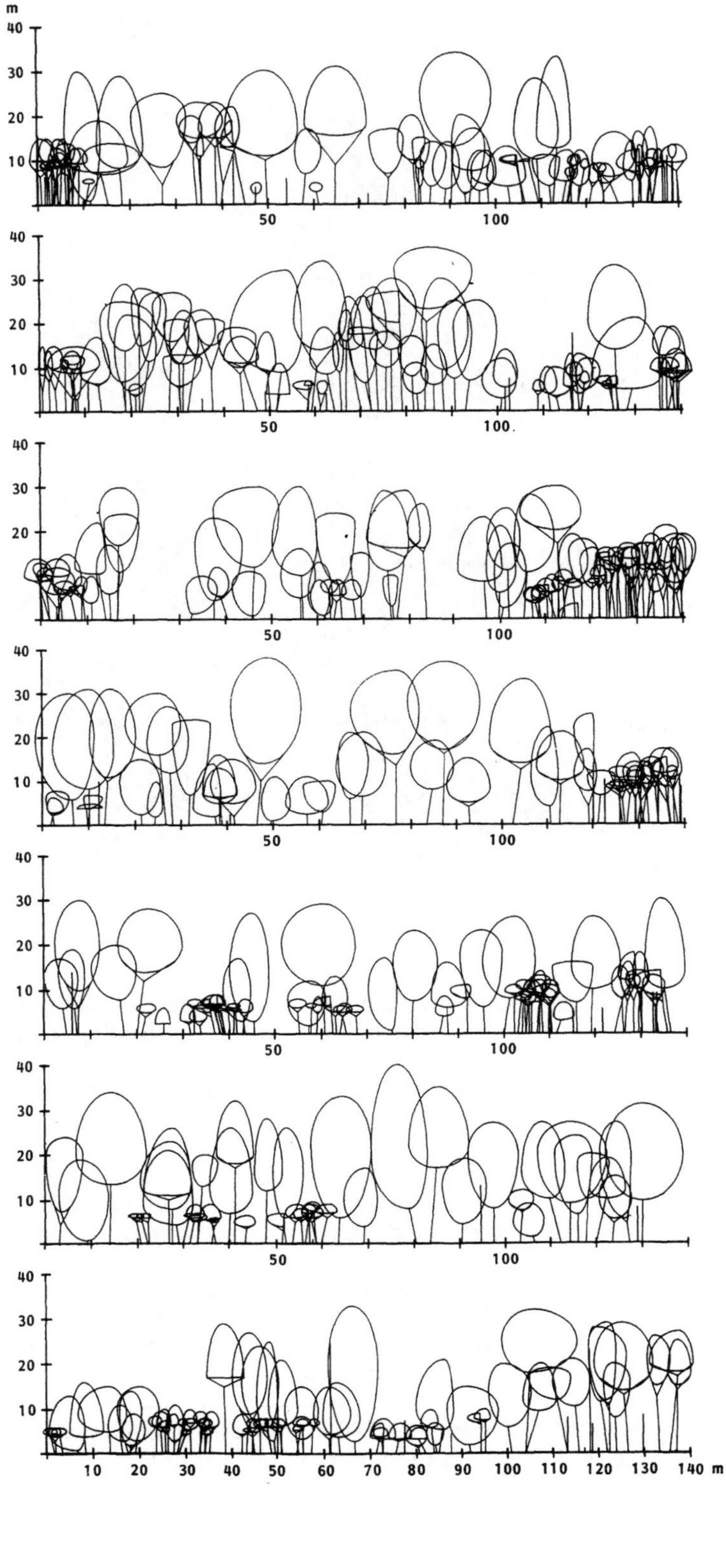

Fig. 7.6.

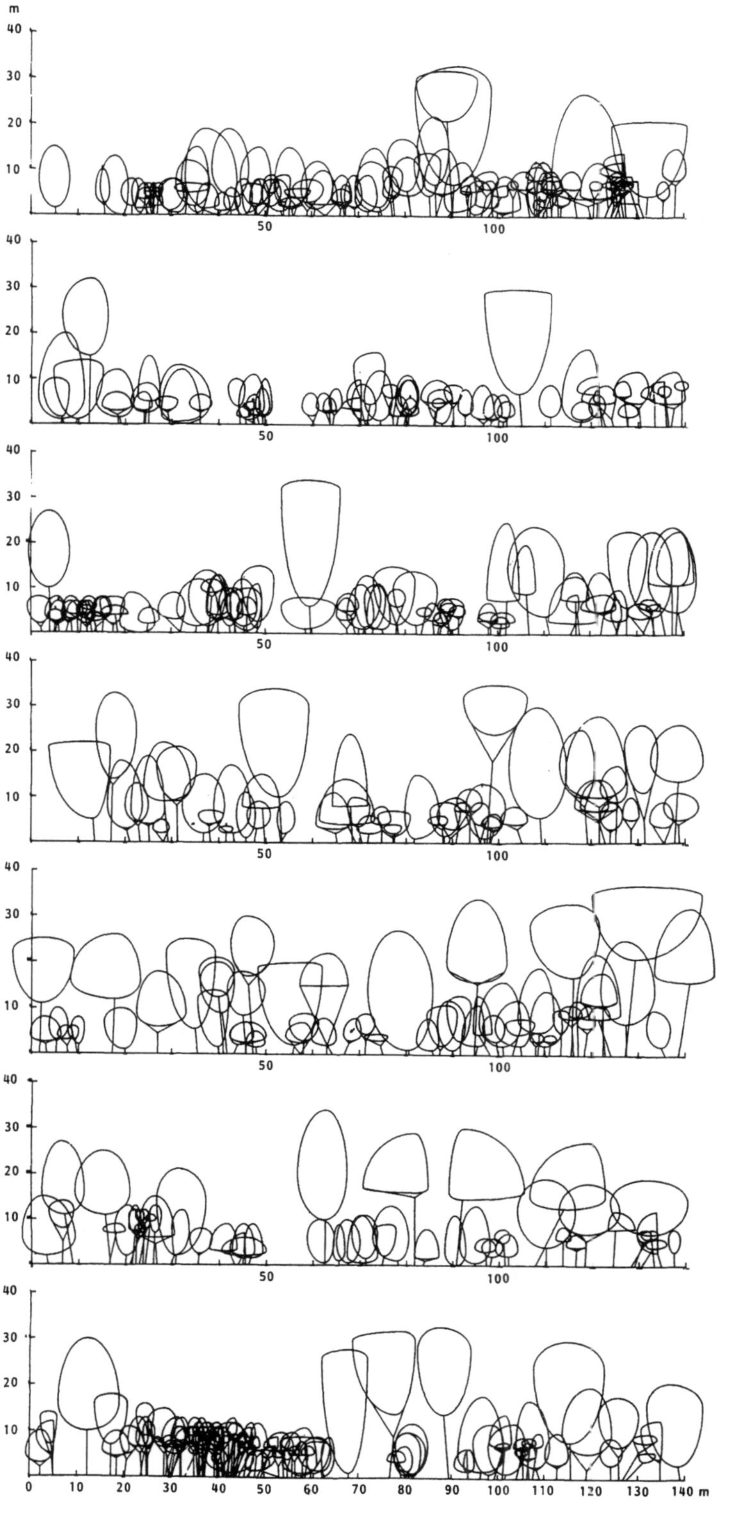

Fig. 7.7.

Fig. 7.7. Side views of seven adjacent 10 x 140 m transects through the core area II in La Tillaie (Fontainebleau) plotted with the program ARBOPLOT

The network pattern is most pronounced in slowly fragmentating biostatic phases. Against the background of the spread of regeneration waves in time and space, of the fact that after 100 years height differences have been bridged between trees of both generations and that the biostatic phase may last for 200 years, it seems likely that the above described network pattern regeneration may have a continuous nature in time.

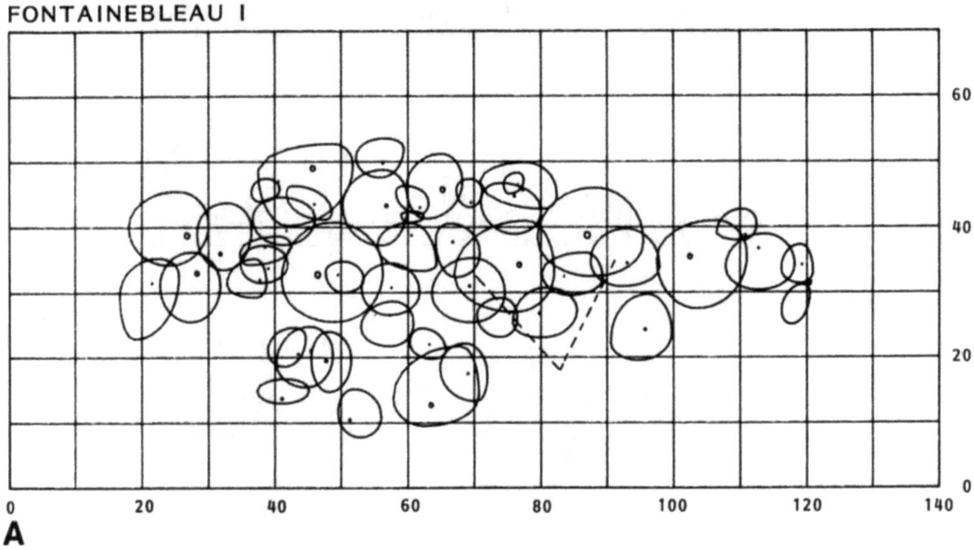

Fig. 7.8. A Computer-plotted ground plan drawing of the area sampled for tree rings in the core area I in La Tillaie (Fontainebleau). **B** Photographs (June 1986) of the sample area with little regeneration under the first generation beech. The position the photographs were taken is indicated in **A**

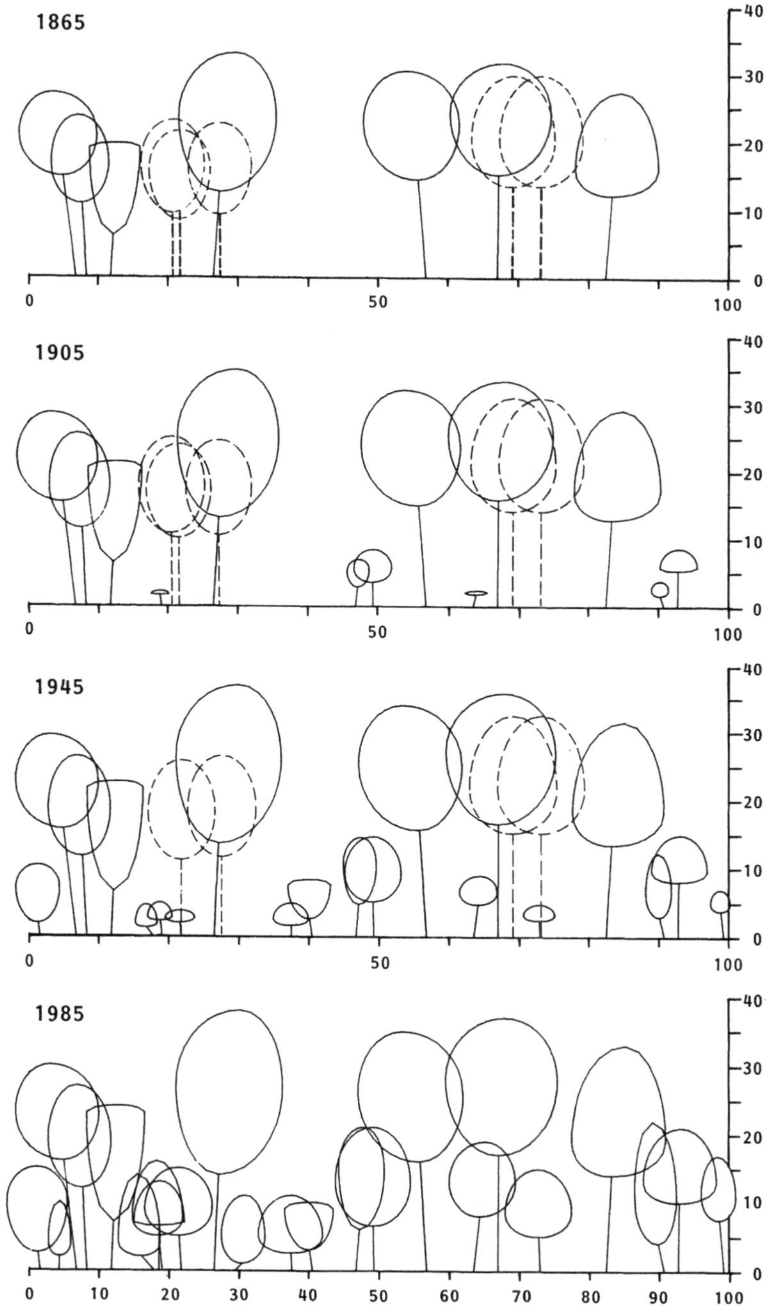

Fig. 7.9. Recontructed forest development using the program GROWRING (see Chap. 6.5). Woody transect area in Core area I of La Tillaie (Fontainebleau). Four pictures at 40 years' intervals have been cut out of the cartoon film that covers 300 years. *Dashed* trees have been reconstructed from tree bodies on the forest floor

7.1.4 Forest Growth Reconstruction in a *Melico-Fagetum* (Core area I)

The traces of uprooting in the northwest of the sample block and east of the transect at 95 m can be dated with the trees that grow on the mound (Figs. 7.8 and 7.9). These trees established themselves in 1880 and 1888 respectively. Trees in the direct surroundings of the uprooting became established in 1875 and 1896 and neighbouring old trees show an increased growth rate after 1880. Probably, these uprootings and the two others in the centre of the core area are of the same age, although the trees in the canopy gaps and on the mounds are not exactly of the same age. It is likely, even though the uprootings occurred simultaneously, that not all trees became established immediately after uprooting nor did they all surmount 1.3 m, at which the tree was cored, in the same year. The tree just west of the uprooted tree in the centre of the core area shows no growth reaction although it must have been released. This could be explained by sudden exposure or damage by the fall of its neighbouring tree. It is likely that these uprootings and other dying trees (which left no trace) triggered the regeneration wave starting in 1850 or released saplings that already waited underneath the canopy.

Three recent uprootings in the central woody transect area, of which the tree bodies still remain, can be dated back to 1940 and 1955 by pin-pointing the growth reaction of surrounding trees. Uprooted trees were shown to have more effect on the growth of surrounding trees than broken-off trees. Therefore the break-off of a tree cannot be dated with the same accuracy as uprootings. Probably the trees had already died when they broke down. A release caused by a tree dying upright is a much more gradual one than in the case of uprooting. One of the trees that broke off can be dated back to about 1940 and two to about 1955.

7.1.5 Forest Growth Reconstruction in a *Fago-Quercetum* (Core area II)

In the woody transect area, two regeneration units can be distinguished (Figs. 7.10 and 7.11, see also Fig. 7.5). The first, at 20-50 m, developed from uneven-aged regeneration of *F. sylvatica*, starting in the southern edge of the canopy gap in 1869. The other trees dominating this unit established themselves during the period 1883-1892. The earlier appearance of trees in the southern edge of the gap may be due to earlier tree establishment and reaching 1.30 m height more quickly because of better growth. Later establishment and less vigorous growth in the northern edge of the gap can be explained by the direct sun irradiance that can lead to desiccation (Pontailler 1979). Two trees now suppressed by the others became established latest, i.e. in 1898 and 1900. From the centre of the regeneration unit to its periphery, tree growth has been less and thus a lens-shaped regeneration unit developed. It seems that since establishment of the unit it had hardly been damaged by the fall of older trees.

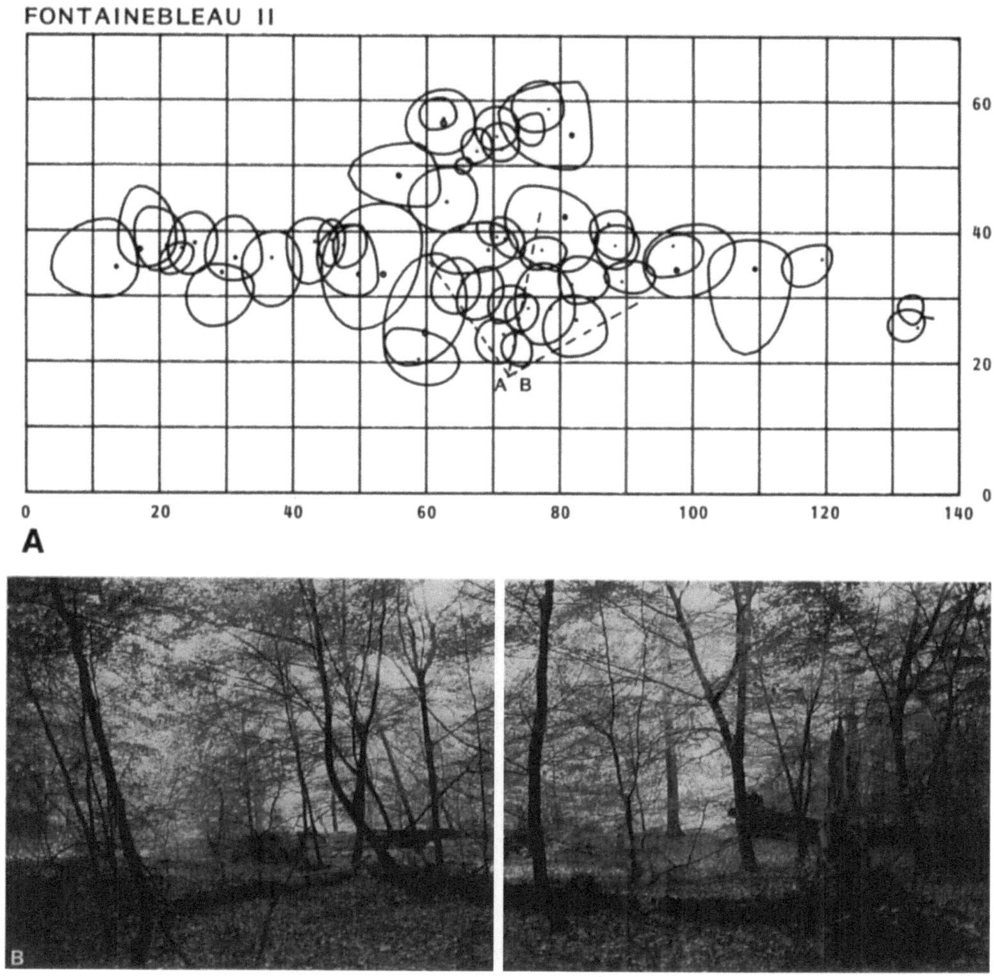

Fig. 7.10. A Computer-plotted ground plan drawing of the area sampled for tree rings in thecore area II in La Tillaie (Fontainebleau). **B** Photographs (June 1986) of the regeneration unit in the centre of the core area, the position the photographs was taken is indicated in **A**

The origin of the canopy gap, below which this regeneration unit developed, is indicated by some slight circular depressions about 50 cm in diameter. These depressions could be the result of decaying stumps of trees that died standing. As the tree base and the last roots decayed, the soil surface caved in for about 20 cm. The ongoing process has been observed elsewhere in Denny Wood (New Forest, England) in beech trees that died because of the 1976 drought. Four trees in the centre of the regeneration unit show traces of accelerated growth since 1940, which might have been caused by the uprooting of three trees west of the transect. Of one of these fallen trees, a beech, some wood remains in stage 5 and 4 of decay, which fits in well with the estimate of 40 years before total decay of big beech trees (Koop 1981). The mound-and-pit micro-relief of the uprootings has flattened to stage 3 (see Chap. 3).

The two oldest trees in the second regeneration unit in the centre of the core area (Fig. 7.10) date from 1883. Other trees became established in 1886, two in 1890 and still others in 1894, 1899, 1900, 1913, 1914 and 1918. In the period 1943-1952 some trees became established in the north- eastern edge of the regeneration unit. This probably

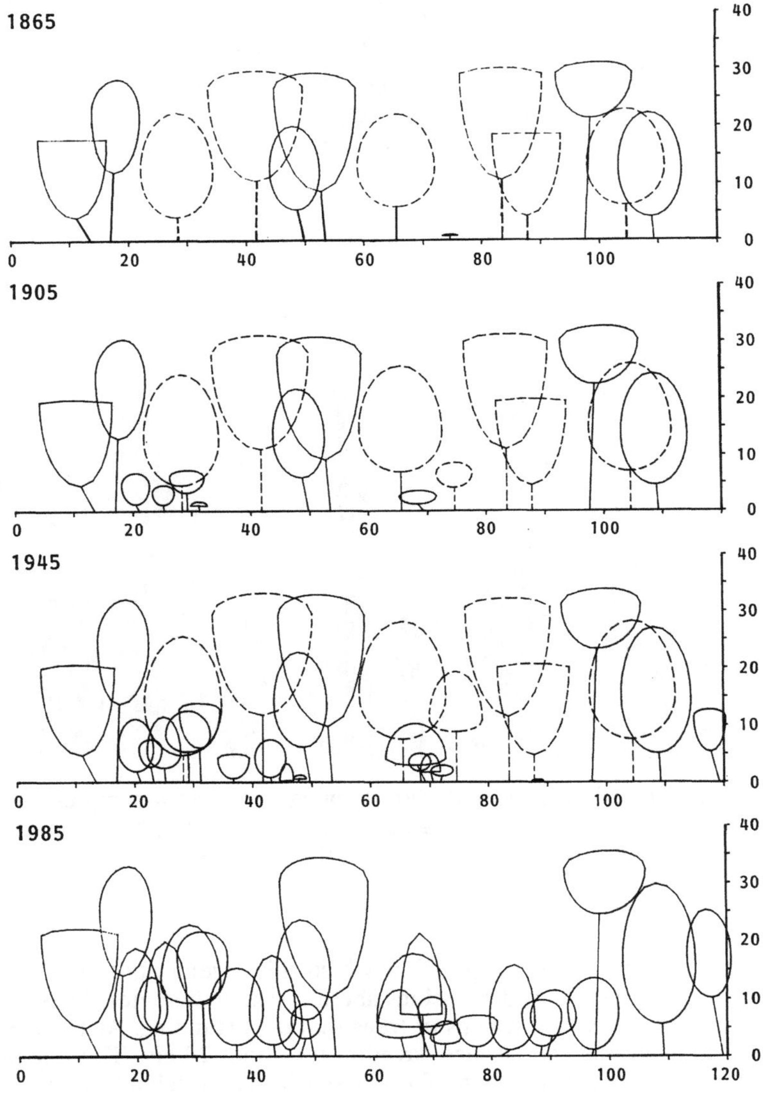

Fig. 7.11. Recontructed forest development using the program GROWRING (see Chap. 6.5). Woody transect area in Core area II of La Tillaie (Fontainebleau). Four pictures at 40 years' intervals have been cut out of the cartoon film that covers 300 years. *Dashed* trees have been reconstructed from tree bodies on the forest floor

happened because of the death of the biggest tree in the regeneration group and some uprootings east of the transect. The origin of the canopy gap is only indicated by one slight circular depression and one trace of uprooting (stage 4). Other depressions or traces of uprooting may have been overlooked under the dense crowns of fallen trees. This regeneration unit suffered from damage caused by falling trees during and shortly after the storms of 1967, as is documented by maps (Bouchon et al. 1973). Where two crowns of big beeches struck the ground, part of the older trees of the regeneration unit probably was demolished. One of the trees became severely damaged, but survived. The unit has been split up into two smaller ones. In the western part of the former regeneration unit tree growth was least vigorous and trees became overgrown by others.

During the storms of 1967 and in the following years, more than ten big trees in the transect and its direct surroundings were uprooted. As a reaction to their release, the surviving trees in both regeneration units show an increased diameter increment after 1967. The old canopy trees also reacted to the release, except for some of the oldest, already senescent, trees. Along the fallen logs of uprooted trees in 1967 younger beeches became established. They were too thin to be bored for tree rings. The up to 3-m-long etiolated sprouts of *Pteridium aquilinum* found in 1980 give evidence that *P. aquilinum* must have occupied part of the canopy gap after the storm of 1967 and became outshaded by the spreading tree crowns of the beeches. Probably because of the shadow cast by the *P. aquilinum* vegetation, the already existing regeneration and the piled-up uprooted tree stems and crowns, no other more light-demanding tree species became established. In its northern end, the transect contains a small part of a third regeneration unit.

7.1.6 Mosaic Development During Regeneration After Management

In the past, the forest reserves have been influenced to some extent by forest grazing by deer and cattle. Cattle grazing lasted until the start of the 20th century. The number of cattle reached its summit under Louis XIV (1661-1715). Apparently this period of high grazing pressure did not hinder the establishment of the second generation of trees (Fig. 7.12). Allowing the inhabitants to collect dead wood and litter was practised until the 19th century, obviously without have caused great damage to the forest (Tendron 1983). With Tendron (1983) we can conclude that the forest reserves developed more or less undisturbed by human interference, at least without cuttings, since 1372. The 200- to 300-year-old trees described by Barrilon d'Amoncourt (1664) must originate from the period shortly after the clear-cut of 1372. The beeches may have spontaneously invaded a pioneer oak forest somewhat later and therefore may have been younger.

Barrilon d'Amoncourt (1664) described the uprooting of beech on the podsol in Petite Tillas (core area II) caused by some heavy storm shortly before 1664, which the oaks survived. He concluded that the beech was not going to survive, which allows the conclusion that the storm must have felled the majority of the beeches. However, he mentioned that second-generation beech became established at the same time. This partial decline of the forest, with oaks surviving, was followed by beech regeneration. The establishment of this tree generation can be dated with tree ring analyses to between 1609 and 1723. Possibly the oldest beeches of this tree generation may have survived the storm as young trees under the canopy of oak.

La Grande Tillas (core area I) is described in 1664 as having a higher portion of beech, which might be explained by a site quality that favours beech in competition with oak. The difference in site quality of core area I and II enables the beeches to become taller than in core area I (Koop and Hilgen 1987). The description of 1664 mentions only "Some parts are rather open because of the fall of old beeches and some wind-thrown oaks". A major decline is reported not earlier than 1716. The range of establishment of the following tree generation from 1651 to 1693, as derived from the tree ring samples, is not as wide as in core area II. It suggests a later but quicker decline, that might be explained by a more or less pure beech forest with a low number of oaks that could survive the decline. The old descriptions give a good explanation of why the present old beeches in core area II are older and show a wider range in their years of establishment than in core area I.

Both more or less even-aged stands that became established after the cut of 1372, decayed after they had reached their physiological maximum age as described by Barrilon d'Amoncourt (1664). The periods of decay took less than one half and one century in core area I and II respectively. The effects of the cuttings in the fourteenth century can still be seen in the present forest structure. The second phase of decline of the almost pure beech forests started after 1850. Until the present day, in both plots part of the old generation has survived and regeneration continued, which means that this second phase of decline has lasted for a longer time than the first one in the seventeenth century. The generation change has spread in time and in space. Koop and Hilgen (1987) reconstructed from distribution of dead wood and stages of uprootings that until 1967 the rate of tree mortality was highest in the *Melico-Fagetum* (core area I). The storms of 1967, however, accelerated the decline in the *Fago-Quercetum* while core area I was hardly affected (Figs. 7.10 and 7.11). The effect of the 1967 storm in the *Fago-Quercetum* resembles the one described on the same site type in 1664.

Korpel (1987) describes regeneration cycles in five natural beech forests in Slovakia that have been studied for more than 20 years (Fig. 7.12a). "The natural beech forest during its approximately 250-270-year-long developmental cycle is subject to substantially smaller changes from the viewpoint of numbers of trees, volume stock and stand structure in comparison with natural forests of other plant communities. Even on a smaller area it is characterized by a considerable diameter differentiation. An expressionless single-layer, height-levelled structure takes place in the optimum stage only on small areas during a short period. The decomposition takes place gradually by way of natural mortality of separate standing trees. It is characterized by an expressive small-scaled texture of the forest" (Korpel 1987).

Niemann (1968), Brünig (1981) and Koop (1986a) describe phase transition in formerly managed woodlands, left unexploited after imposing the strict forest reserve status. Niemann (1968) characterized the mosaic development as follows. Initially, in the designation phase, forest structure is still determined by former management. In comparison to the state of equilibrium, the designation phase is characterized by an incomplete mosaic of developmental phases. Few phases, with man-made boundaries, occupy a large share of the mosaic. Depending on how far forest structure differed from a state of equilibrium at the moment of reserve designation, a phase of collapse may intervene before an equilibrium has established. The so-called initial phase starts after collapse of the managed stand. It may last for several tree generations until, through spontaneous fragmentation, a complete mosaic of phases, characteristic for the state of equilibrium as described by Korpel (1987), has become established (Niemann 1968; Brünig 1981). Koop (1986a) presented a model for the repeated wave-like generation changes after management that is supported by the data presented here (Fig. 7.12b). The state of equilibrium may be reached in a fairly short time when forest stands, characterized by a wide variation in developmental phases in a heterogeneous mosaic, are selected and designated to be reserves (Prusa 1985).

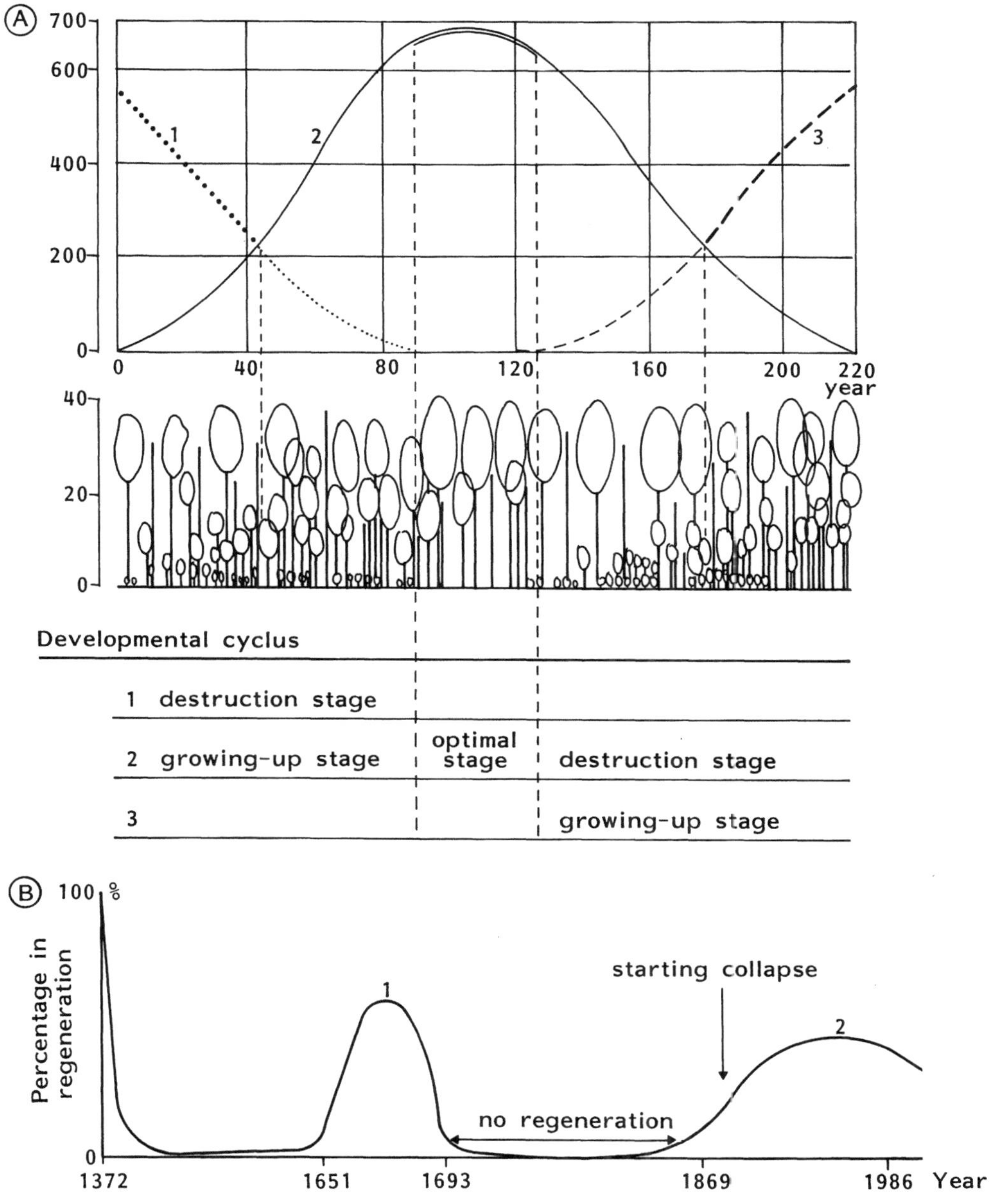

Fig. 7.12. A Model showing the developmental cycle of beech forests after Korpel (1981, 1987). Repetition of mutually overlapping developmental cycles (1 to 3) represented in changing stand structure and stand volume, after data from Slovakian virgin beech forests. **B** Model after Koop (1986a) of surface percentage of a regeneration mosaic being in decay against time as reconstructed from the core area I in the Fontainebleau forest reserve

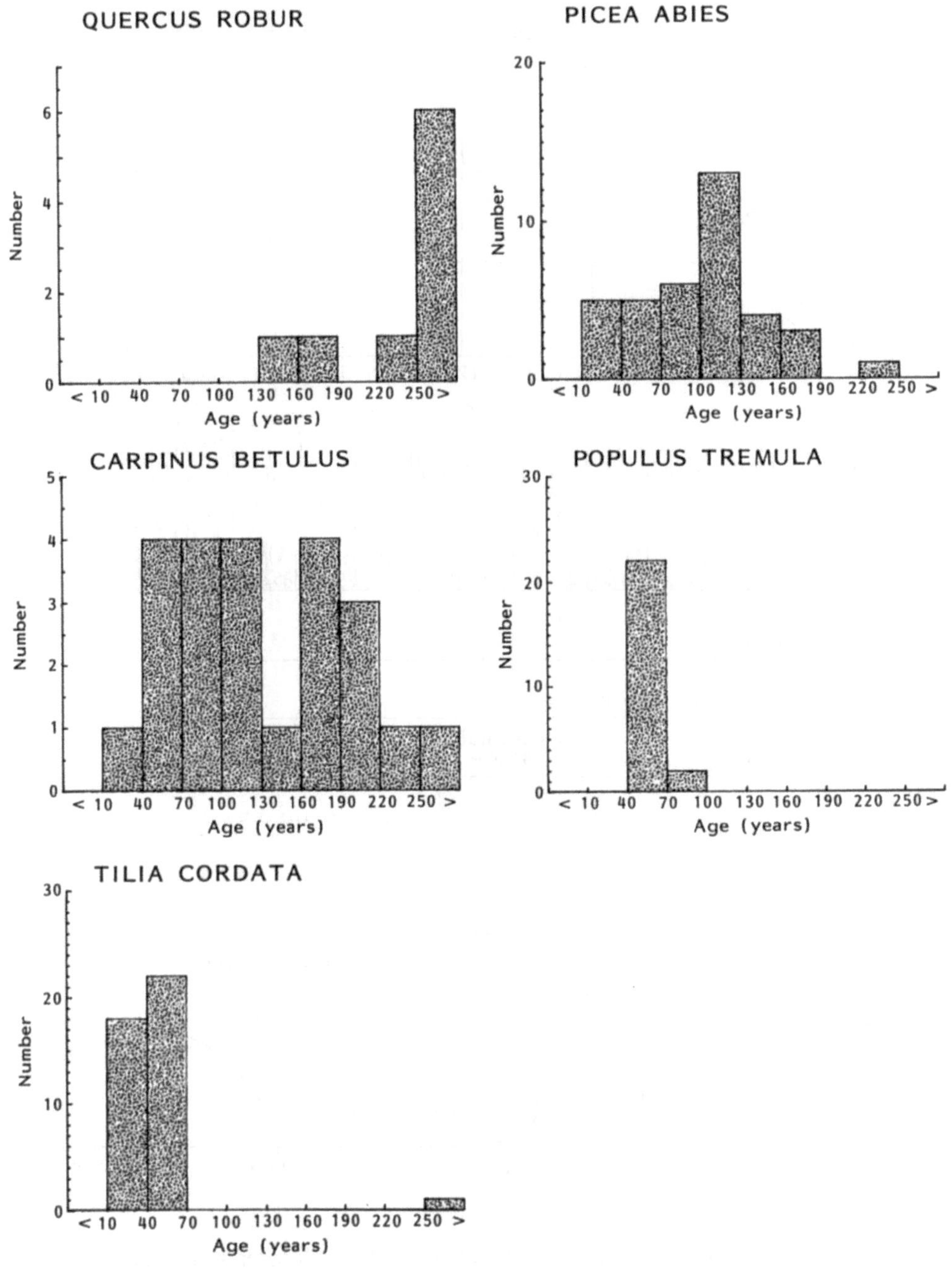

Fig. 7.13. Age distribution diagrams of *Quercus robur*, *Picea abies*, *Carpinus betulus*, *Populus tremula* and *Tilia cordata*

7.2 Cyclic Forest Dynamics in Unexploited Lime-Hornbeam Forest in Bialowieza (Poland)

The National Park of Bialowieza is part of the 1250 km^2 Bialowieza forest on Polish and Russian territory situated between 52°.30' and 52°.57' northern latitude and 23°.20' and 24°.20' eastern longitude. The climate is of a continental subboreal type, with long snow cover in winter of average 92 days and a relatively short vegetation period of 185 days. Mean annual precipitation amounts 641 mm (Falinski 1986).

In 1977, four 1-ha plots were established in the *Tilio-Carpinetum*, the vegetation type most closely related to the West-European *Stellario-Carpinetum* (Derkman and Koop 1977). The hectare blocks established in 1977 were mapped for developmental phases. They showed a small-scale regeneration mosaic. Trees were measured in circular sample plots within the hectare blocks. A clear relation between tree density and mean tree diameter was found in the sample plots. Differences in tree densities in thicket phases could be distinguished between released thicket phases that established under a closed canopy and newly established thicket phases after canopy gap formation. Regeneration of *Quercus robur* only was found in one plot in a bigger gap. Regeneration of *Ulmus glabra* preferentially occurred in dense *Urtica dioica* vegetation. An 200-year-old clear-cut could still be recognized in present forest structure. A wavelike regeneration of *C. betulus* and *T. cordata* followed the pioneer stand of *P. tremula*, *B. pendula*, *Q. robur* and *P. abies* after decay (Koop 1986a).

In 1982 and 1987, two of these plots in compartment 256 and 342 were reassessed according the methods described in Part A. The core area in dep. 342 coincides with the I.B.P. area measured earlier by Falinski (1966). In 1984, vegetation mapping in both core areas was carried out by Van de Bos (1984). Tree ring samples of all trees with a Dbh > 10 cm in a central 40 x 50 m block in the core area in compartment 256 were taken in 1987 by Bevers and Blokland. Besides, all *Populus tremula* trees in the rest of the core area were bored. The core area of compartment 256 discussed here belongs to a *Tilio-Carpinetum* (Rijken 1976). The parent material is of glacial origin. The soil conditions of the *Tilio-Carpinetum* are described by Falinski (1986). The typical brown earths are differentiated by gleying and lessivage, resulting in pseudogley soils and pseudopodsols. The core area is situated on a pseudogley soil in sands upon boulder clay. In a deeper sand cover a pseudopodsol is found in the northern edge of the core area carying a more acid local *Trientalis* variant of the *Tilio-Carpinetum*.

7.2.1 Historical Analysis

After it became came under Polish rule as a result of union of the Great Duchy of Lithuania in the fifteenth century, the Bialowieza forest was a hunting field of the Lithuanian dukes and Polish kings. There was a special guard to watch over the forest. Exploitation for timber remained backward because the Baltic forests were more open for transport overseas. After the third partition of Poland in 1795, the forest passed under Russian rule and was not as well protected by law as in the three centuries before. The forest became exploited for timber and the stock of game diminished. A great fire in 1811, and a year later the invasion of Napoleon's troops devastated the forest. Successive surveys in the period 1861-1890 indicate the continuous shrinking of the forest area. After 1893 the Bialowieza forest was ranked as a private hunting-field of the Russian Czar, and a new castle was built in Bialowieza. Exploitation for timber was

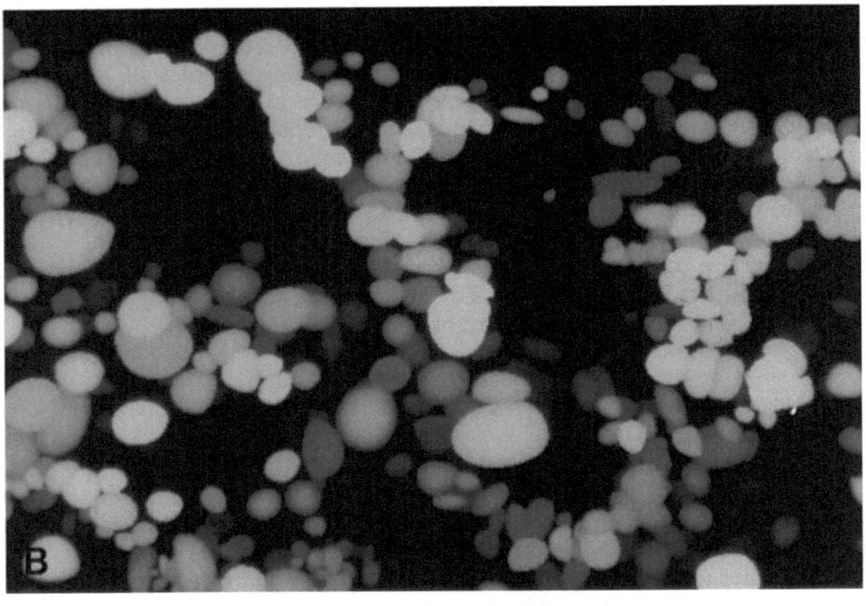

Fig. 7.14. Canopy height computed from the forest model in the core area compartment 256 in the National Park of Bialowieza (Poland). **A** Canopy height of all trees, showing a fragmented canopy of emergent trees. **B** Canopy height of potential trees, clearly showing the pattern of the network of aggrading regeneration units surrounding the emerging trees

prohibited and great quantities of game were let free, which caused an extremely high stock. During the period 1892-1915 the forest actually acted as a game-breeding park, which resulted in destruction of the forest undergrowth, debarking of young trees and inhibition of rejuvenation of most broadleaved trees. As a result, *Picea abies* spread under the canopy of deciduous trees in nearly all habitats (Falinski 1988). Maps of Karvec (1903) and of the forest service of 1958 (both in Falinski 1986) show the increase of stands with prevailing spruce (*Picea abies*). During the German occupation in World War I, hunting without control decimated the game, while 5,000,000 m^3 of wood were cut and a network of narrow gauge lines was built for removing it (Falinski 1986). Gruber and Voit (1917) and Escherich (1928) write about fires in the forest lighted by farmers and Russian convicts. Most fires started on clearings with *Picea* pruning rests. After Poland became independent in 1921, the central part (47 km^2) of the forest was designated as the oldest Polish National Park. During World War II, Bialowieza was ranked as a hunting area. In 1947 the Bialowieza National Park statute was finally approved.

7.2.2 Tree Age Distribution

Of the eight *Quercus robur* trees sampled for tree rings, three were established in the seventeenth century, three in the eighteenth and two in the nineteenth century. In competition with *C. betulus* and *T. cordata*, *Q. robur* has little chance in this core area. In another core area (compartment 371) in a bigger gap the species has established itself successfully.

Only few live *P. abies* trees older than 120 years can be found. *P. abies* reinforced its predominant position by a regeneration wave during the period 1874-1906 (Fig. 7.13). This increase of *P. abies* was clearly described by Paczoski (1930). It might be explained by the enhanced influence of game during Russian management which favoured regeneration of *P. abies* against other tree species (Falinski 1986, 1988).

The tree ring measurements of two *C. betulus* trees, now dated back to 1896 and 1900, are not very reliable. The combination of faint tree rings typical for the species with periods of slow increment gave some problems. The heart of these trees was rotted so the tree age could only be determined by extrapolation. *C. betulus* establishment shows two gaps, one from 1824 until 1862 and the other from 1900 until 1923. The latter may have been caused by the high grazing pressure in those days (Falinski 1986, 1988). Except for one 344-year-old *T. cordata* tree, all trees became established more or less continuously after 1922. The increase of *T. cordata* regeneration is similar to the findings of Wloczewski (1972) after 33 years observation of the permanent plot compartment 319. All *Populus tremula* trees established themselves between 1918 and 1938.

7.2.3 Pattern in the Regeneration Mosaic

Similar to the core areas of the beech forests of Fontainebleau, in the core area of Bialowieza a network pattern can be recognized (Figs. 7.14 and 7.15). While in Fontainebleau a height limit of 20 m is used to separate the emergent trees from the potential set, in Bialowieza - because of more than one species that reach different maximum heights - the whole set of potential trees can be depicted (Fig. 7.14b).

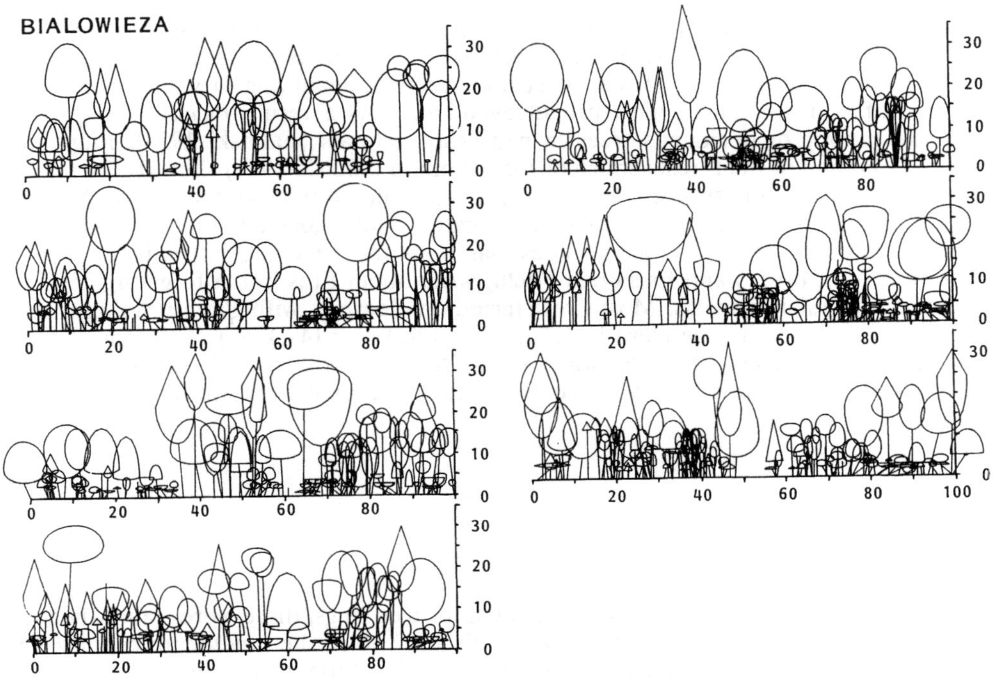

Fig. 7.15. Side views of seven adjacent 10 x 100 m transects through the core area in compartment 256 in the Bialowieza National Park plotted with the program ARBOPLOT

The highest trees in this regeneration network occur in more or less coherent regeneration units.

7.2.4 Forest Growth Reconstruction

Traces of uprooting and the positions of dead wood in the core area (Fig. 7.16) belonging to the *Tilio-Carpinetum* (Falinski 1986) bring to the fore some information about forest dynamics in the far past. Uprooting is a common forest process. Many authors reported on it, especially in regions with remnants of virgin forests or where forest floors had not been levelled (Lutz 1940; Stephens 1956; Armson and Fessenden 1973; Stone 1975; Brown 1976; Koop 1981; Beatty 1981). Brewer and Merritt (1978) and Beatty (1981) considered the area of disturbed soil of an uprooting to be half an ellipse. Hutnik (1952), Lyford and Maclean (1966) and Putz (1983) reported on the influence of uprootings upon tree distribution. Londo (1976), Beatty (1981) and Koop (1981) described the influence on species composition in the herb layer. Oliver and Stephens (1977) and Zeide (1981) describe methods of mound dating by research on trees that grow on them.

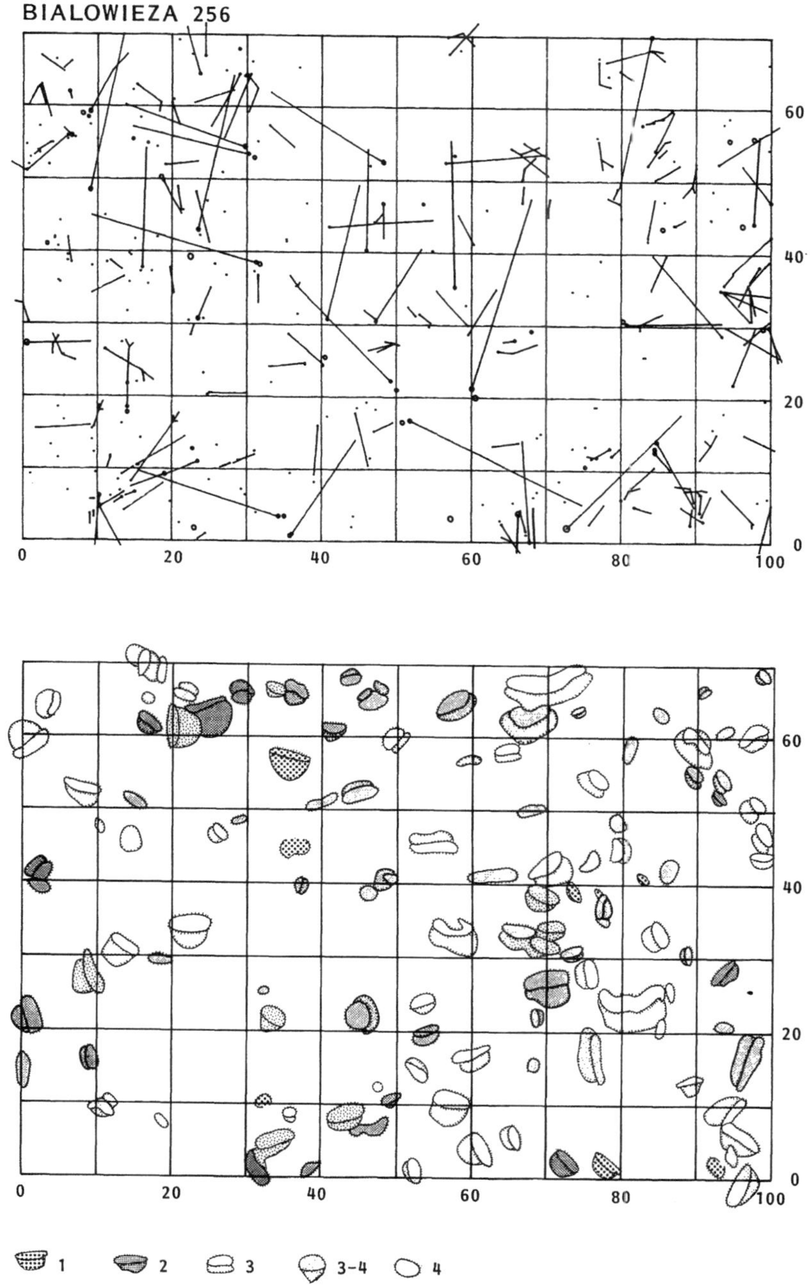

Fig. 7.16. A Lying and standing dead trees in the core area in compartment 256 of the Bialowieza National Park. **B** Traces of uprooting in different stages of flattening (see *numbers*) in the core area in compartment 256 of the Bialowieza National Park

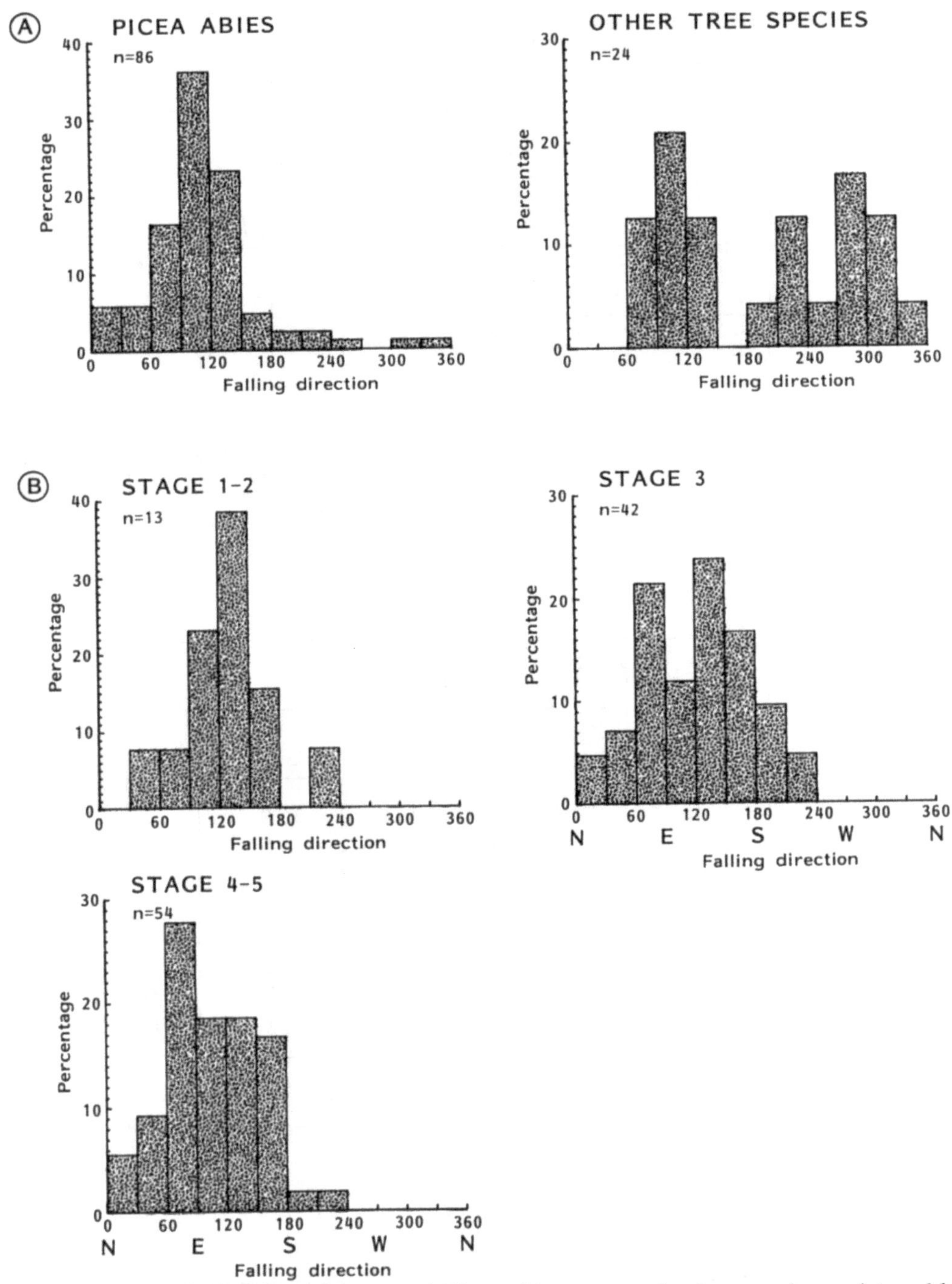

Fig. 7.17. A Fall direction of uprooted *Picea abies* trees and other species registered by Falinski (1976) in the autumn of 1971, on a 100-ha surface of the *Tilio-Carpinetum*. **B** Fall direction of uprooted trees in stage of flattening 1-2, 3 and 4-5, interpreted from the mound-and-pit micro-relief, as being perpendicular to the longitudinal axis of the mound away from the pit in the core area in compartment 256 of the Bialowieza National Park

Henry and Swan (1974) used the information provided by the position and state of decay of dead wood for reconstructing the forest history of a forest on a restricted area. Falinski (1977, 1986) showed that repeated charting of dead wood can be a useful tool for detailed studies on dead wood dynamics. Mortality, translocation and decomposition of wood, as an important part of the forest ecosystem flow of organic material, has been studied by Falinski (1977) and Sollins (1981).

The direction of tree fall of an ancient uprooting can be read from the mound-and-pit micro-relief, as being perpendicular to the longitudinal axis of the mound away from the pit. However, the sequence of mound and pit may be interchanged in the case of uprooted trees falling against other trees. Because of a return stroke the mass of earth clinging to the root system is moved to the other side of the pit, causing a reflected image of the normally formed mound-and-pit micro-relief (Koop 1981; Beatty 1981). The falling direction may then be misinterpreted by 180°. In recent uprootings in the core area, this effect was only observed in deciduous trees and not in *P. abies*.

Nevertheless, Falinski (1976, 1986) noted that trees obviously really fall against the wind. The author concludes that for the fall of the tree, not the direct wind slashes but indirect wind turbulations might be responsible. Kuiper and Van Schoten (1985) and Kuiper (in preparation) working with Douglas fir, stress the importance of swinging and oscillations as a cause of uprooting.

The direction of ancient tree falls interpreted from mound-and-pit micro-relief in the core area shows a huge peak from north-east to south-east (Fig. 7.17). Tree falls registered by Falinski (1976) in the autumn of 1971, on a 100-ha surface of the *Tilio-Carpinetum*, show a similar peak, but only for *P. abies* (Fig. 7.17).

For deciduous tree species, the absolute numbers of uprooted trees are much lower than for *P. abies*, 24 against 86. Moreover, their direction of fall is more spread (Fig. 7.17). The peak around 280° and its inverse around 100° ($280\text{-}180^{o}$) might be explained by the above-mentioned processes of flyback and fall against the wind. In conclusion, it seems that *P. abies* was the species most often uprooted in the core area. This conclusion is substantiated by the form and dimensions of the mounds and pits. Recent *P. abies* falls caused very long, low, stretched mounds with half-circular, shallow pits. Deciduous trees cause a much more compact micro-relief, i.e. an elliptical higher mound and a smaller but relatively deeper pit. In the oldest uprootings (stage of flattening 3 and 4), some immensely long mounds are found, while the shallow pits can hardly be recognized.

Nine *P. tremula* trees grow on the mounds of uprootings of stage 3 and another 14 in their direct surroundings. All these trees established between 1918 and 1938 (Figs. 7.16, 7.18 and 7.19b). Growth reactions of trees in the direct surroundings of the uprooting indicate that all uprootings of stage 3 can be attributed to one storm around 1917. Two types of growth reactions indicate the date of uprooting, i.e. short negative growth reactions especially of the undergrowth *P. abies* trees, caused by damage of falling trees and lasting positive, slowly falling, growth reactions of seven *Carpinus betulus* trees caused by release. After 70 years, no remnants of the uprooted trees remain. The *P. tremula* trees are arranged in clearly separate regeneration units. Possibly their quick clustered establishment was favoured by their ability of clonal rejuvenation (Koop 1987).

Besides trees being uprooted, many old trees, especially *P. abies*, broke down without uprooting. Many lying tree bodies and tree stumps in a progressed stage of decay (code 5, see figure III.6e) were mapped (Fig. 7.16). Probably these trees died before they fell. Using growth reactions upon release of surrounding trees, the death of these trees could be dated from 1939 to 1944. The oldest woody remains of *P. abies* therefore are

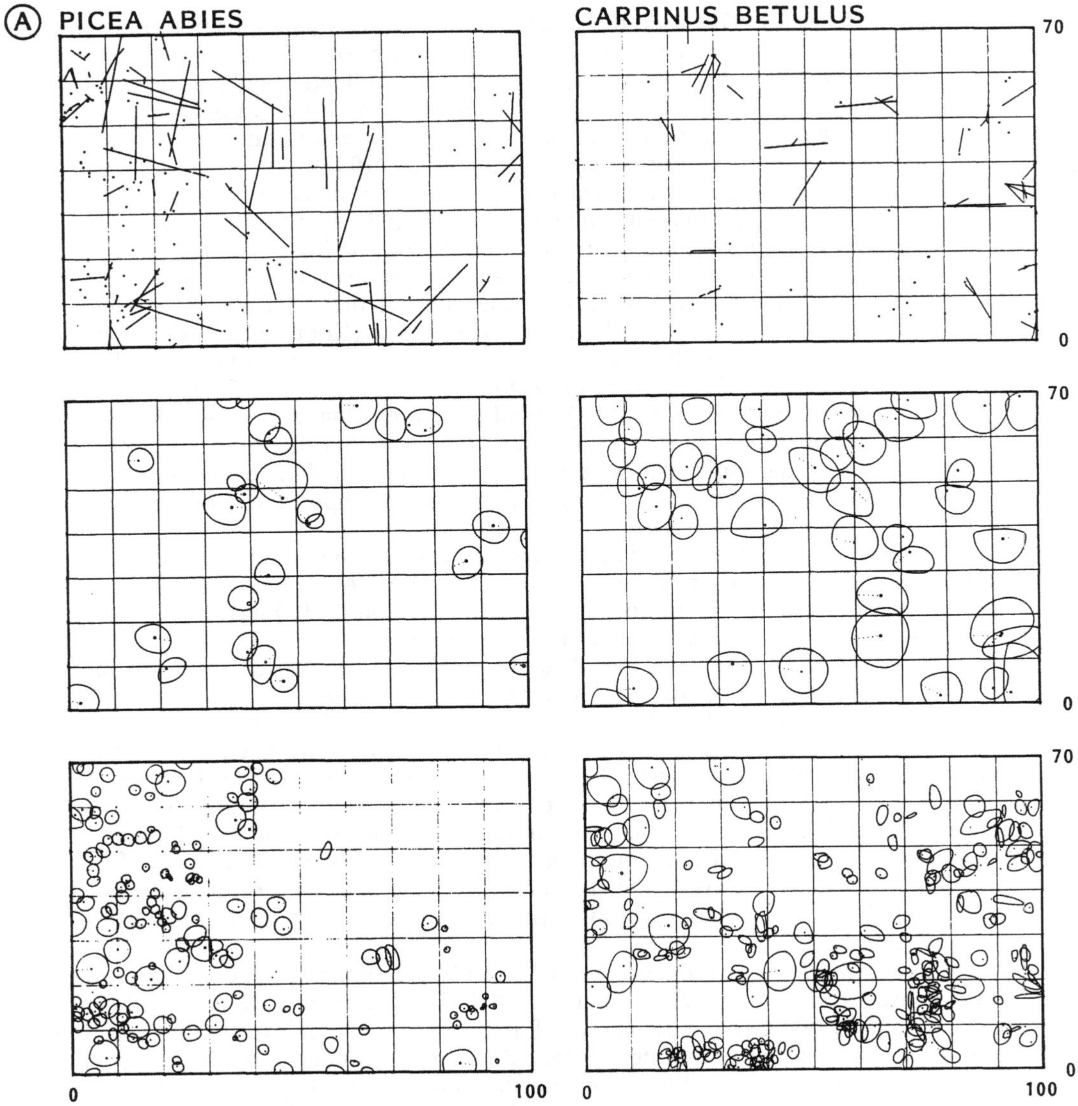

Fig. 7.18. A Computer-plotted ground plan of, respectively, dead trees, trees of the present and potential trees of *Picea abies*, *Carpinus betulus*. **B** (see next page) Computer-plotted ground plan of *Quercus robur*, *Populus tremula*, *Corylus avellana* and *Tilia cordata* in the core area of section 256 of the National Park of Bialowieza

not older than 50 years. The fall sometimes can be reconstructed by short negative growth reactions of trees damaged by the fall 10 to 15 years after the death of the tree. This period of declining *P. abies* probably triggered the establishment of the second regeneration wave of *P. abies*. Some *T. cordata* trees show increased diameter increment after 1947 and many trees became newly established during the following years (Fig. 7.13). In 1984 and 1985, two storms threw many *P. abies* and some *C. betulus*. These uprootings belong to stage 1 of flattening (Fig. 7.16). Only five uprootings of stage 2

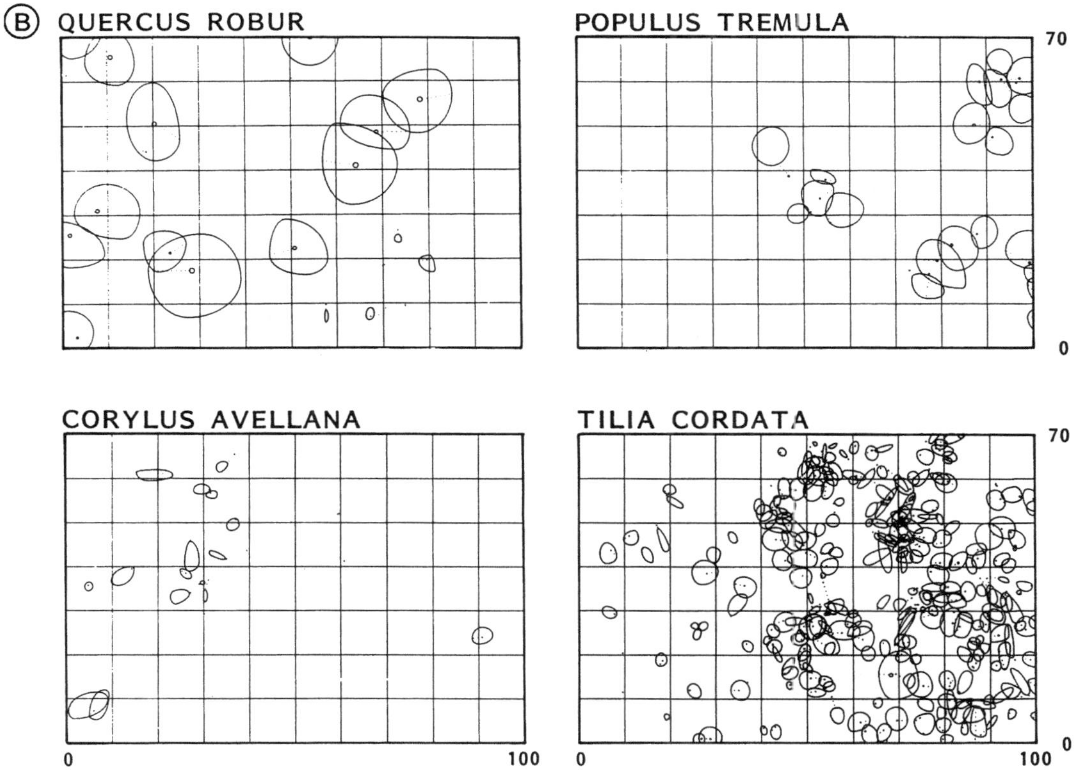

could be mapped, one of which could be dated back to 1955. The others are situated outside the area where tree ring samples were taken.

The oldest stage of flattening, i.e. stage 4, could not be directly dated like stage 3 except in the case of one 353-year-old *Q. robur* that grows on one of the oldest uprootings. Synchronous periods of growth increase or decrease in old *Q. robur*, *P. abies* and *C. betulus* trees can indicate local release or canopy closure. Some old trees show increased growth in 1755 and 1810 and a decrease of their growth in 1775. The traces of uprooting in stage 4 of flattening probably belong to the oldest period of growth increase. Some of the intermediate mound-and-pit micro-relief between stage 3 and 4 might be dated back to 1810. However, these data, coming from only a few trees, cannot help to reconstruct an early forest growth.

During the period 1860-1900 all trees show a lasting decrease in growth, which might be explained by canopy closure of the *P. abies*-dominated forest. Possibly soil compaction and modified nutrient turnover during the period of high grazing pressure might have had some effect on tree growth as well. The gradual growth increase around 1900 might be explained by a natural thinning and starting decline of the 17th century *P. abies*, generation together with a backlog of regeneration due to the high grazing pressure, which gave the remaining trees more room.

During the eighteenth and nineteenth centuries, big *P. abies* trees dominated a large part of the transect area (Fig. 7.20). Some parts were dominated by *C. betulus* and a big *Q. robur*. Under the canopy of these broadleaved trees, *P. abies* trees became established at the end of the eighteenth century. Some of these *P. abies* trees have now overgrown the *C. betulus* canopy. In 1917, like in the rest of the core area, many big *P. abies* trees were uprooted. This event triggered the establishment of three groups of *P. tremula* trees (Fig. 7.18), of which one group together with an *Acer platanoides* tree, can

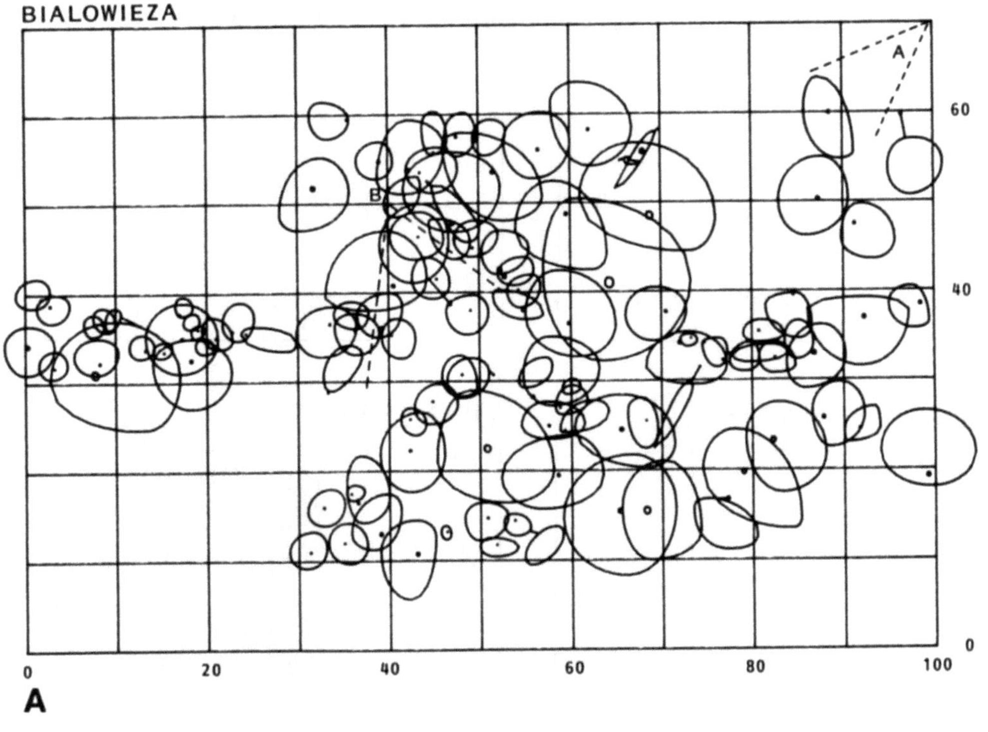
BIALOWIEZA
A
B
60
40
0
0
20
40
60
80
100
A

B

be found in the middle of the transect. During the natural thinning in *P. abies* during the late 1940's, in the first part of the transect a new generation of *P. abies* became established. Until 1984 a gradual decline of *P. abies* and *P. tremula* took place. The canopy gap formed by the death of an old *C. betulus* in 1943 was filled by a regeneration unit of mainly *T. cordata*. The young *T. cordata* trees at the margin of the opening show lateral crown growth in the direction of the light. Hallé et al. (1978) referred to this phenomenon as the "river-bank effect". The storms of 1984 and 1985 uprooted the remaining emergent *P. abies* trees.

7.2.5 Mosaic Development During Regeneration

Areas with many living trees or dead tree bodies of *P. abies* and areas with high numbers of older uprooting traces without traces of a tree body of *P. abies* almost mutually exclude each other (cf. Figs. 7.16, 7.18). This means that *P. abies* once dominated the whole core area but drew back successively since 1917 from different parts of the core area. The dominant role of *P. abies* is confirmed by the map of stands with prevailing spruce by Karcev (1903). In 1982 only three of these very old trees survived. Many of the tree bodies on the forest floor, as well as those trees that were uprooted about 1917 and could be traced by the mound-and-pit micro-relief, probably belonged to this generation. Apparently about 1917 uprooting played an important role, while in the 1940's many trees died while standing. The recent storms of 1984 and 1985 uprooted and broke many *P. abies* trees including the last ones of the eighteenth century generation. *P. abies* is the most shade-tolerant species in this area and it can withstand periods of severe grazing pressure better than the other species. The oldest tree sampled for tree rings in the core area is 228 years old. This suggests that the species is an opportunist that could reach a dominating position during the eighteenth century. The establishment of the old trees that died and were uprooted in the period 1900-1945 cannot be explained by the high stock of game during Russian management, although in this period the *P. abies* regeneration was relatively favoured (Falinski 1986, 1988) (Fig. 7.13). Conditions in the eighteenth century must have favoured the old generation of *P. abies*. However, what these conditions actually were, could not be reconstructed from the data. The age structure of *P. abies* with regeneration periods in the eighteenth, late nineteenth centuries and around 1945 shows some similarity with the wavelike tree generations of the beeches in Fontainebleau.

C. betulus, which had a maximum age in the core area of 264 years, shows an almost continuous establishment. There are two interruptions, the first during canopy closure of *P. abies* in the nineteenth century and the second during high grazing pressure in the period 1900-1923.

The oldest regeneration units in the network of potential trees probably originate from the 1917 storm. The part of the network that crosses the transect at 80 m (Fig. 7.20) dates back to 1942.

Fig. 7.19. A Computer-plotted ground plan drawing of the area sampled for tree rings in the core area of compartment 256 of the Bialowieza National Park (Poland). **B** Uprooting mound-and-pit micro-relief of stage of flattening 3 with *Populus tremula* trees on the mounds (indicated with *white cross*) in core area compartment 256 in the National Park of Bialowieza June 1987. Position of the photograph indicated in **A**

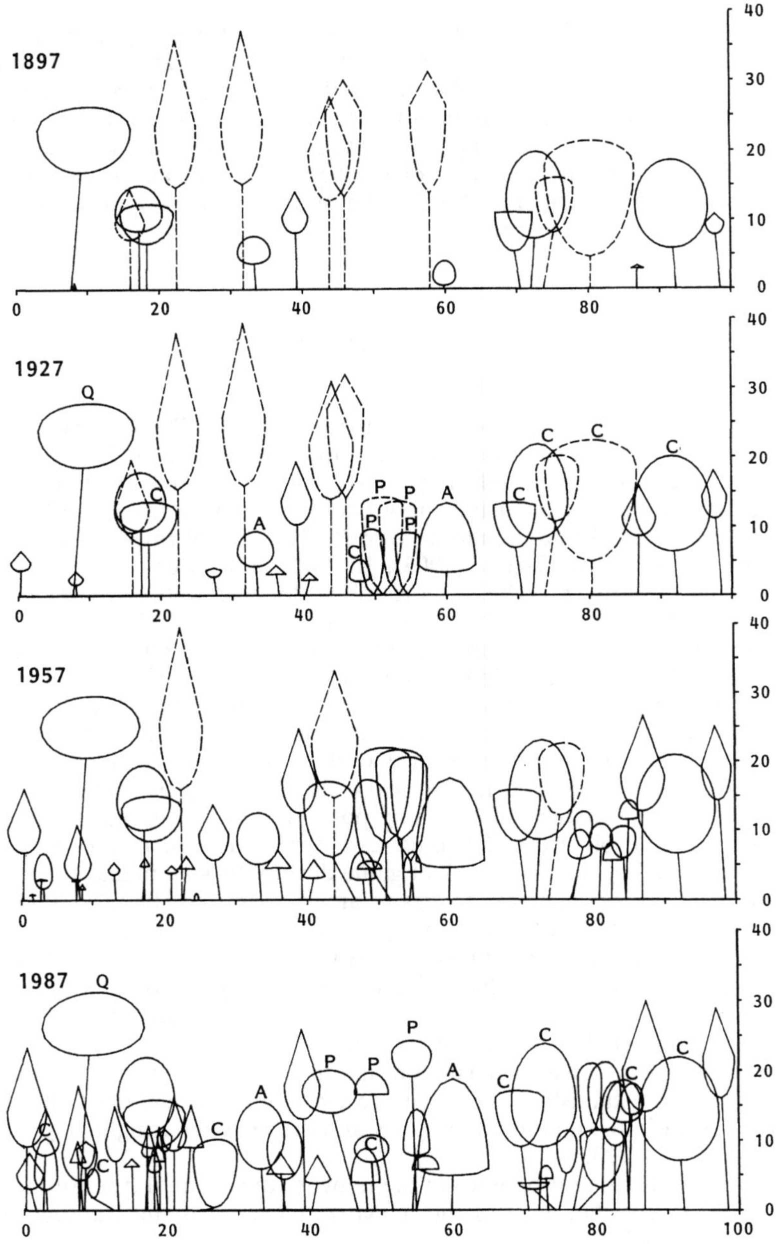

Fig. 7.20. Reconstructed structural forest development using the program GROWRING (see Chap. 6.5). Woody transect area in core area of the section 256 in the National Park of Bialowieza. Four pictures at 30 years' intervals have been cut out of the cartoon film that covers 200 years. *Q Quercus robur*, *C Carpinus betulus*, *P Populus tremula*, *A Acer platanoides* other trees *Picea abies* and *Tilia cordata*

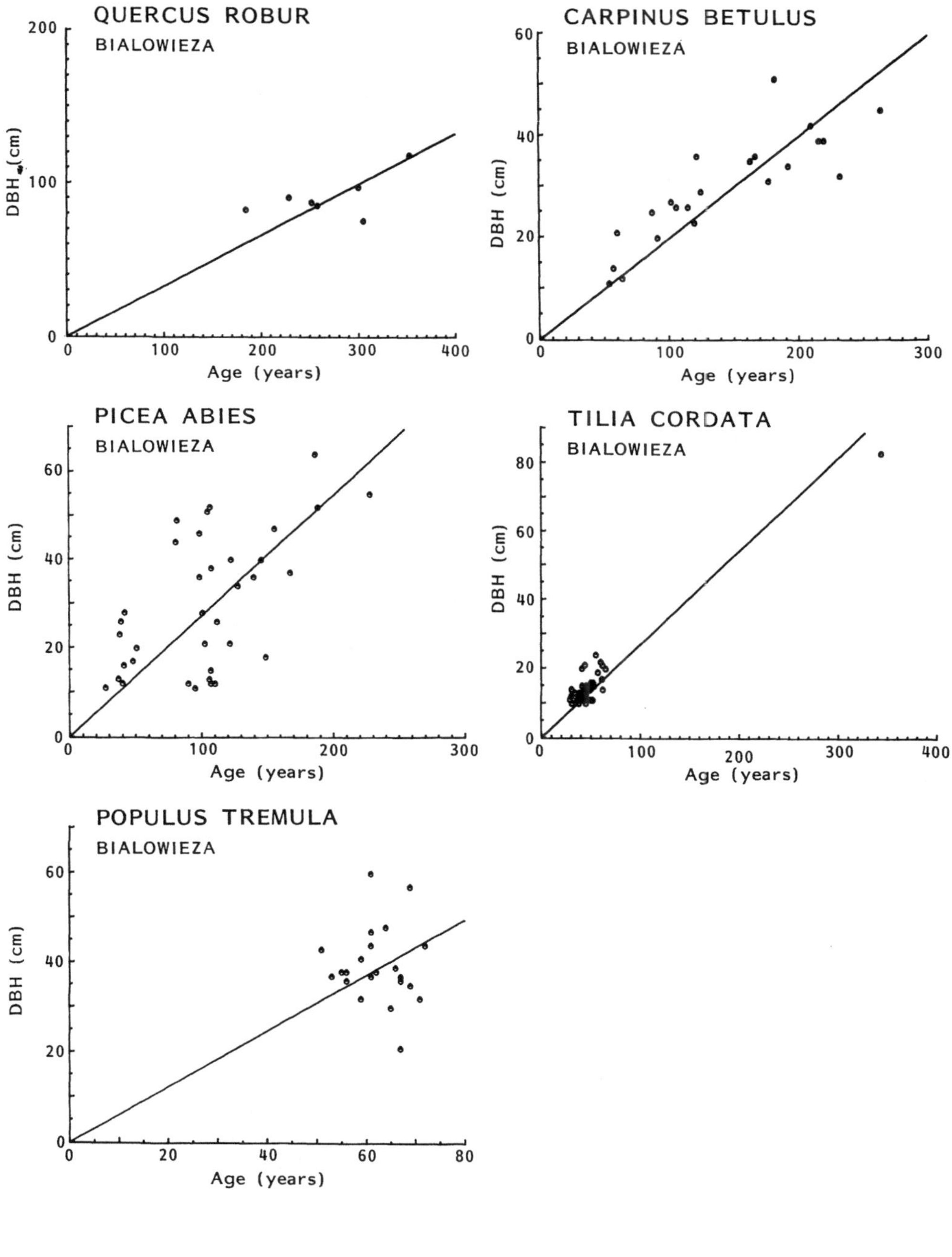

Fig. 7.21. Relation between tree age and Dbh for all trees sampled for tree rings in the core area of compartment 256 of the Bialowieza National Park

The story of mosaic development is more complex than in Fontainebleau because of different tree species, each with their own growth rates and maximum life-span. In Figure 7.21 the different Dbh increment rates with age per species are plotted. For example, a *P. abies* tree of 100 years old can have a Dhh of 10 cm or of 50 cm. The tree species can wait for long time periods in the shade and then accelerate growth. Canopy gaps filled by *P. tremula* have turned into a tree phase of more than 20 meter in height within 40 years. Nevertheless, their life-span will be between 70 and, say, 100 years. On the other hand mature *Q. robur* and *T. cordata* trees grow slowly and have inhabited parts of the silvatic mosaic for centuries. The oldest *T. cordata* was 344 years old.

7.3 Reconstruction of Forest Dynamics in Pasture Woodlands in the New Forest (England)

In the New Forest in 1980 the first three woody transect areas were laid out. Because the effects of forest grazing upon forest structure have their main impact upon shifts in the silvatic mosaic, in the design of nested plots the level of the core area has been left out. Instead, more and longer woody transect areas have been analyzed to capture all aspects of forest grazing. In 1983, Drenth and Oosterbaan (1984) enlarged the number of transects to ten with a total length of 1.73 km. In 1988, the transects were assessed for the second time. The reconstruction of forest dynamics will be different from the methods used in the two previously described forest reserves. It will be based on tree ring analyses of other authors and on interpretation of actual patterns of forest vegetation and recent developments.

7.3.1 Historical Analysis

Peterken and Tubbs (1965) discuss the regeneration phases in the unenclosed woods of the New Forest. Although regeneration has always been present to some degree, tree ring countings in 24 woods confirm distinct tree generations. The generations can be linked with cuttings or a fall in grazing pressure. There is a Pre-A-generation of which only 24 oaks throughout the whole forest have been recorded.

The A-generation dates from 1625-1685. The 1707 survey of the New Forest records the very heavy fellings that went on during the reigns of Charles I and Charles II (Flower 1980). Flower mentions the notes accompanying the 1707 survey which say that: "All the above 12,000 trees, suitable for the Navy, may be felled over the next 40 years without detriment to the Forest, there being so many young trees, not yet of a sufficient size". The generation of young oaks referred to in the notes must have dated from around the middle of the seventeenth century and coincided with the A-generation. The existence of this generation during the Stuart period is confirmed by Presentments of the Regards in 1660. In regard to the nature of these fellings there is direct evidence for preferential removal of oak for Navy shipbuilding purposes. Ridchards et al. (1787 in: Flower, 1980) describe that amongst other woods Rushpole wood has been cleared for oak for the Navy. The wood is described as: "Oak/beech wood with some fine oaks remaining, but most having been felled for the Navy". The A-generation was followed by a relatively short period during which relatively little regeneration took place. In the period 1725-1825, in response to further oak removal for shipbuilding, widespread regeneration took place.

Fig. 7.22. Photograph of the 300-year-old A-generation beech forest of Bratley Wood in one of the transects in the New Forest. While other parts of the Bratley Wood are collapsing gradually (see Fig. 7.24) this part still survived (June 1988)

The central parts of the ancient woods in the New Forest, under study with the information system, such as Denny Wood, Mark Ash Wood, Bratley Wood and Berry Wood, almost purely consist or consisted of the 300-year-old A-generation and its follow-up. They show a rather homogeneous structure of more or less even-aged beech trees with a Dbh around 100 cm (Fig. 7.22). Regeneration units of beech, even if they differ in age, tend to fuse, as in Fontainebleau, to even-sized units. Most of the A-generation trees show a short-stemmed wide-branching habit, with wide growth rings in their youth, all the result of development in open conditions.

The second marked phase of regeneration common to all unenclosed woods, the so-called B-generation, can be dated from 1851-1875. It is a direct result of the rapid fall in grazing pressure brought about by the deer removal act of 1851 (Fig. 7.23). In the transects in the Denny Wood and the Berry Wood, this generation resulted in a woodland encroachment on the edges with oak.

The last or C-generation from 1935-1955, which occurred throughout the forest, was a response to a rapid fall in the numbers of cattle and ponies due to a collapse of the meat market.

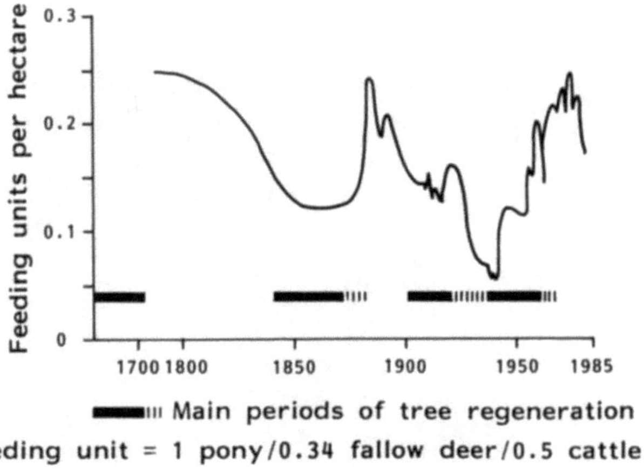

Fig. 7.23. Grazing and browsing pressure and tree regeneration phases. Numbers of animals have been converted into feeding units per hectare (one feeding unit =1 pony =0.5 cattle =0.34 fallow deer).(After Tubbs 1986)

Fig. 7.24. Photograph of the collapsing 300 years old Bratley Wood in the New Forest (June 1988)

7.3.2 Pattern in the Regeneration Mosaic

Generation changes after a phase of decay, such as described for Fontainebleau, can only succeed if grazing pressure allows young trees to emerge from the browse line. Under governing grazing pressure in Fontainebleau, the generation changes take place without any intervening treeless phase, or at least within 15 years the gap was reoccupied by young trees. It depends upon the size of more or less even-aged forest patches and their differences in age to what extent a phase of decay will occur and how long a total turnover of a forest mosaic will take. After two spontaneous generation changes following the cutting in Fontainebleau, the phase of decline spread in space and time. Without grazing pressure, which delays forest regeneration, an even-aged stand tends to fragment into a mosaic of uneven-aged patches. Large-scale cuttings, such as those in 1372 in Fontainebleau and of the seventeenth century in the New Forest, countered the tendency to spontaneous fragmentation.

If grazing pressure is high enough to delay tree regeneration, the decline in old forest patches will change into a short treeless vegetation. How long gaps will remain treeless depends on a drop in grazing pressure, brought about by changed use of the grazed area as a whole or by a decline in total numbers of grazing animals. If for longer periods regeneration is retarded, collapsed forest patches fuse and open up the forest. Then after a sudden drop in grazing pressure, a wavelike regeneration in the open forest structure is triggered, as happened in the New Forest in 1851 and 1935. Especially trees of a pioneer character as *Quercus robur* and *Betula spp.* take advantage of the open. In beech-dominated *Fago-Quercetum* forest, *Pteridium aquilinum* plays a role in competition for light with beech decreases. In the New Forest in the Bratley and Berry

Woods such spread of *P. aquilinum* could be observed after the initial collapse of the old beech-dominated central parts of the woods (Fig. 7.24). In other places the same process was observed when clear-cuts are practised or if the forests on these sites are transformed into forests with light-transmitting species such as *Pinus sylvestris*, *Quercus spp.* or *Betula spp.*. In permanent quadrats established by Westhoff (unpublished) in clearings in the Weenderbos (The Netherlands), *P. aquilinum* phases exist for 40 years unchanged (Fig. 3.1).

Under their dense canopy and in their loose and accumulating litter layer no other species became established. In the New Forest, *P. aquilinum*, although not eaten, is trampled and its litter is compacted. Grasses and saplings of birch and *Malus spp.* can germinate and thus the duration of a *P. aquilinum* phase can be shortened. In the Denny Wood after the start of the decay phase, triggered by the dry summer of 1976, the 200- to 300-year-old central part of the wood collapsed. During the collapse, which was accelerated by cutting and removal of dead trees, grazing pressure was too high for any regeneration to succeed. Within a few years the forest vegetation changed into a lawn (Fig. 7.25). Meantime the forest on the edge is still expanding into the adjacent heathland.

Fig. 7.25. Photograph of the collapsing centre of the 300-year-old Denny Wood in the New Forest (June 1988). The forest floor became covered by a species-rich lawn

7.3.3 Mosaic Development During Regeneration

Development of a mosaic after the start of grazing depends on the coarseness of the pre-existing mosaic. If a fine-grained mosaic is the starting point, long periods of "overgrazing" retaining generation changes will reduce age differentiation and result in fusion of formerly uneven-aged forest patches into open areas. In general, grazing will retard spontaneous fragmentation of uneven-aged forest stands unless grazing pressure differs at relatively short distance or grazing pressure fluctuates in time. Differentiation in grazing pressure at short distance can be induced by complexity of the environment. Complexity of the environment can be produced by lying dead trees and crown cages that obstruct grazing animals, or by scattered canopy gaps where establishment is triggered and growth of trees is faster than under a homogeneously closed canopy, with few slowly growing young trees. The chance of survival of at least some young trees in such complex environments is greater than in homogeneous environments.

When grazing starts in an open landscape or a clear-cut area, a high grazing pressure will prevent the development of one homogeneous even-aged phase that covers the whole area. While a great deal of young trees are eaten, thorny shrubs can protect trees that establish within their crown. This way of cage regeneration of *Quercus* has been mimicked by forest managers in the past by planting trees together with thorny shrubs. Flower (1980) found the technique mentioned in a document of 1609.

Nevertheless, it depends on the site type what and whether at all shrubby species can establish. Nutrient-poor sites such as *Fago-Quercetum* and *Betulo-Quercetum* (Van der Werf 1989) are beyond the ecological amplitude of species as *Crateagus spp.* and *Prunus spinosa*. To some degree *Ilex aquifolium* can replace their niche at those sites. When grazing pressure fluctuates and time intervals of low grazing pressure are long enough, trees can become established and rise above the browse line. Long time intervals with high pressure will affect the age structure of the forest mosaic. An example of fine grain mosaic of forest patches can be found in some parts of the Rushpole Wood (Fig. 7.26).

All generations, A, B and C, can be found and occasionally regeneration has taken place between them until the present day. An extra artificial fragmentation was brought about by cuttings in 1969, planned to encourage regeneration from a forester's point of view. The operation caused much damage to the glades and rides because of the use of heavy vehicles (Tubbs 1986).

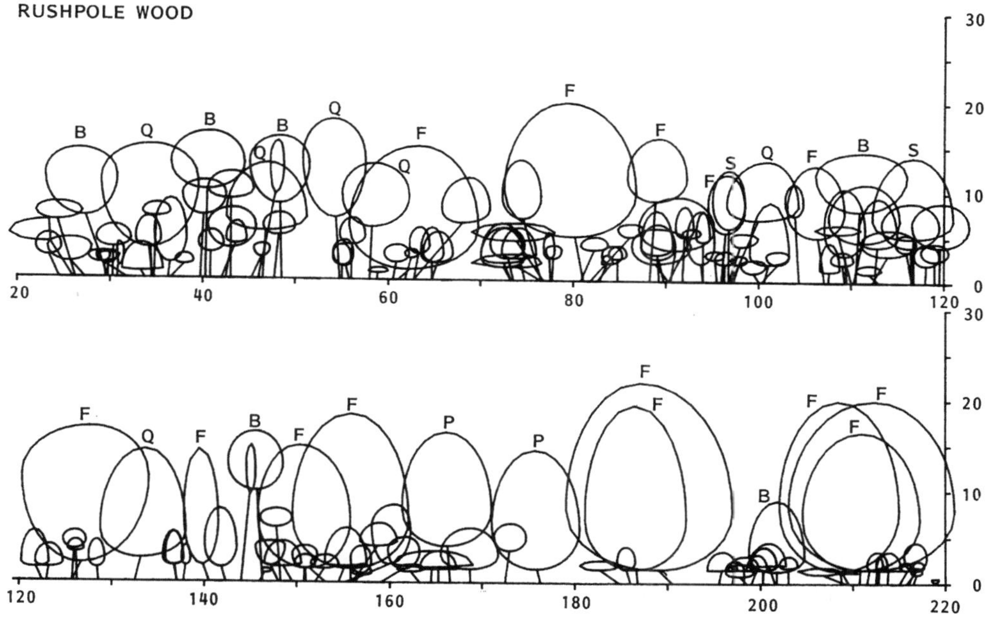

Fig. 7.26. Computer plotted side view of the transect in the Rushpole Wood in The New Forest

Chapter 8

Reconstruction of Long-Term Non-Cyclic Forest Dynamics

In this chapter two examples of successional forest development will be analyzed, one on a sediment alluvial soil and the second on a growing peat soil.

8.1 Non-Cyclic Forest Dynamics in Unexploited River-Bank Willow Forest on the Ile De Rhinau (France)

The present river-bank *Fraxino-Ulmetum* on the Ile de Rhinau along the river Rhine near Strasbourg was sampled for tree rings by Vandeursen and Wisse (1985) in 1981. In 1984 a core area was charted around the transects established by Vandeursen and Wisse in 1981.

8.1.1 Historical Analysis

Although the river has changed its main course several times, the existence of the site where the core area is situated can be documented from old maps, back until at least 1830. On a management map of 1880 the area is characterized as an *Alnus-Salix* forest. Because of canalization of the Rhine since 1840, a narrowed riverbed profile increased the erosive impact upon the forest. Since the "Grand Canal d'Alsace" was built in the period between 1920 and 1970, the erosion and river dynamics have declined. Only when water levels are very high, is the forest flooded (Carbiener and Schnitzler 1987, 1988). The soil profiles under the woody transect area show an undulating gravel bed at a depth fluctuating from 40 to 120 cm. The gravel bed is covered by stratified deposits with varied texture. The lowest groundwater levels vary from 140 to 100 cm. The upper 15 cm consist of biologically homogenized loamy sand.

8.1.2 Tree Age Distribution

The age distribution diagram of all trees of which tree rings were sampled (Fig. 8.1) shows distinct generations. The oldest trees belong to the species *Quercus robur*. They date from 1830 to 1900 with a clear peak around 1890. According to the management map of 1880, at this time the forest was dominated by *Salix* and *Alnus*. The *Salix* and *Alnus* trees probably became established on newly sedimented bare areas,

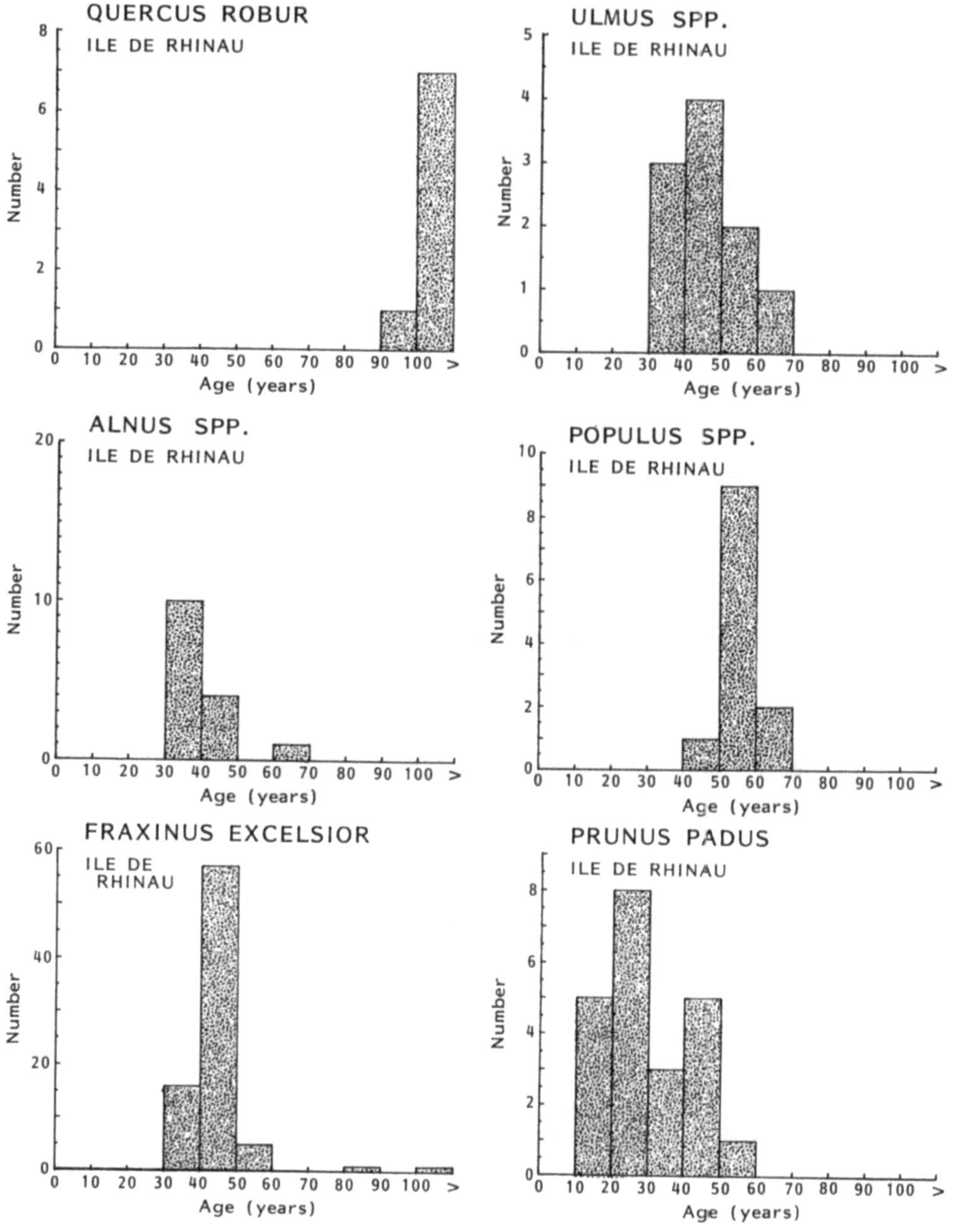

Fig. 8.1. Age distribution diagram of all trees sampled for tree rings in the woody transect area of the Ile de Rhinau. Tree age is established by counting tree rings. If the tree core was not in the sample, the remaining tree ring numbers were estimated as the distance to the tree core divided by the average tree ring width

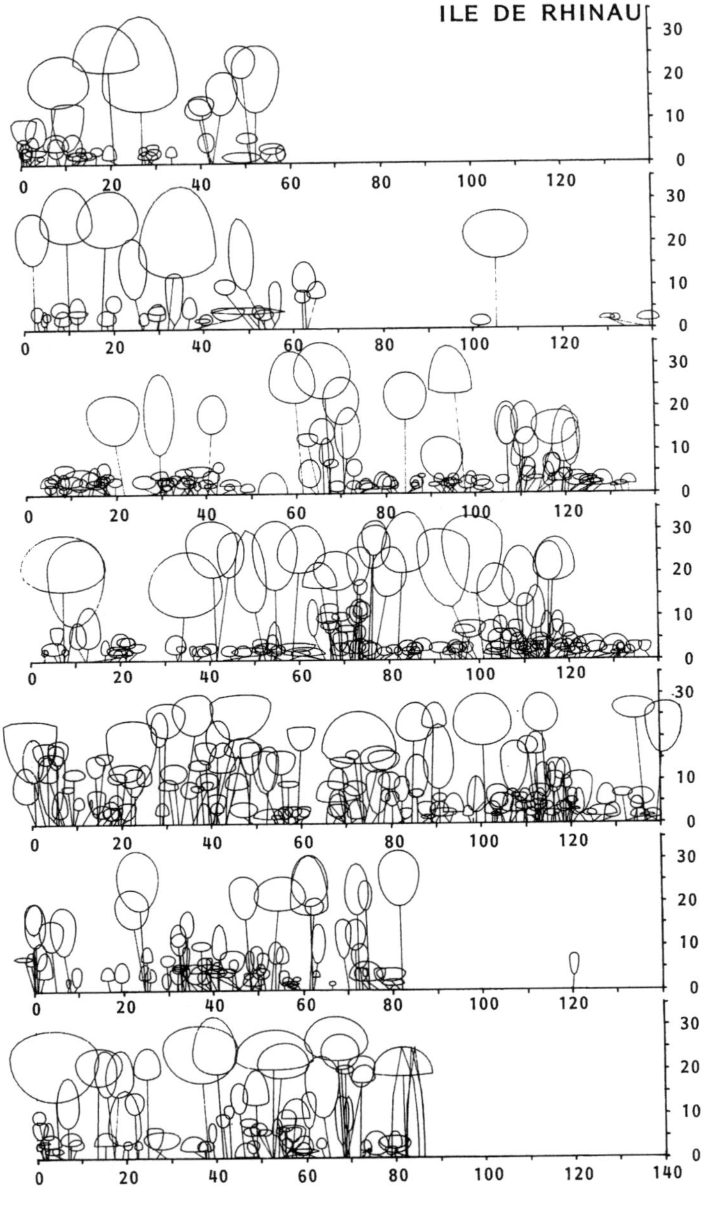

Fig. 8.2. Side views of three 10 x 140 m and four 10 x 80 m transects through the core area in Ile de Rhinau (France) plotted with the program ARBOPLOT

resulting in an even-aged forest. Because of the Dutch elm disease, many *Ulmus* spp. trees died and were cut in 1977. In the transect area only three small stumps of cut *Ulmus* spp. trees are found. Thanks to tree ring counts by Walter (unpublished) on the stumps of the trees cutover a much larger area around the core area, three peaks in the age distribution of *Ulmus* spp. could be distinguished, similar to those of *Q. robur*, a small peak at the beginning of the nineteenth century, a much bigger one around 1890 and the third small one in the years 1910-1930 (Vandeursen and Wisse 1985). In the transect area only trees of the latter period of regeneration are represented (Fig. 8.1).

If we assume that the *Salix-Alnus* forest became established in the same period as the oldest oaks and elms at the beginning of the nineteenth century, it is likely that the establishment of second-generation oaks and elms around 1890 coincides with the starting collapse of the pioneer forest. From the trees established at the beginning of the nineteenth century in the core area only, a few *Q. robur* and one *Fraxinus excelsior* survived.

8.1.3 Pattern in Regeneration Mosaic

Although Van de Winkel (1980, 1984) described network pattern in the river-bank forest Wyhlerwald along the Rhine in the French Alsace, no particular network pattern can be found in the regeration mosaic in the core area (Fig. 8.2). The Wyhlerwald is older in succession to a *Fraxino-Ulmetum* and since floodings belong to the past because of embankment even a further succession can be observed toward a *Carpinetum* (Carbiener et al. 1987).

In the south-eastern edge of the core area at 120 m, a dense blanket of *Clematis vitalba* covers the ground and at the edges climbs into the trees (Fig. 8.3c,d). Beekman (1980) in similar forests in the Alsace found under most *C. vitalba* blankets cut tree stumps. He found that for establishment of *C. vitalba*, canopy gaps of more than one tree height in diameter are necessary and that regeneration of trees must lag behind. According to personal comunication with the forest manager, shortly after 1930 big *Ulmus* spp. were cut in the present core area at the place of the *C. vitalba* blanket. The young trees that establish themselves under the blanket of *C. vitalba* or that become overgrown at the edges of the canopy gap are pulled to the ground by the weight of the lianes. It seems that once the cover of lianes has become established, it is very difficult for trees to establish themselves. In the core area the lianes have prevented trees from establishing for at least 60 years. The observations are too short to draw any conclusion about the increase of the *C. vitalba* cover at the expense of the trees at the forest edge. At 80 m in strip 7 of the core area (Fig. 8.2), some *C. vitalba* stems with diameters up to 12 cm hang down from the canopy of an old *Quercus robur* tree. Because the lianes hang down freely, they must have become established and have grown up together with the oak tree without having prevented the oak tree from growing to its present dimensions.

8.1.4 Forest Growth Reconstruction

Two woody transect areas were sampled for tree rings. Both transects represent a *Fraxino-Ulmetum*. The first transect traverses a former stream gully (between 60 and 70 m) that can be characterized as a *Fraxino-Ulmetum alnetosum* (Van der Werf 1989). Connected with the first transect the second follows the stream ridge at the edge of the same gully. Although both transects form an angle of about 45° (Fig. 8.3), in the side view the second transect has been rotated to be in line with the first one (Fig. 8.4).

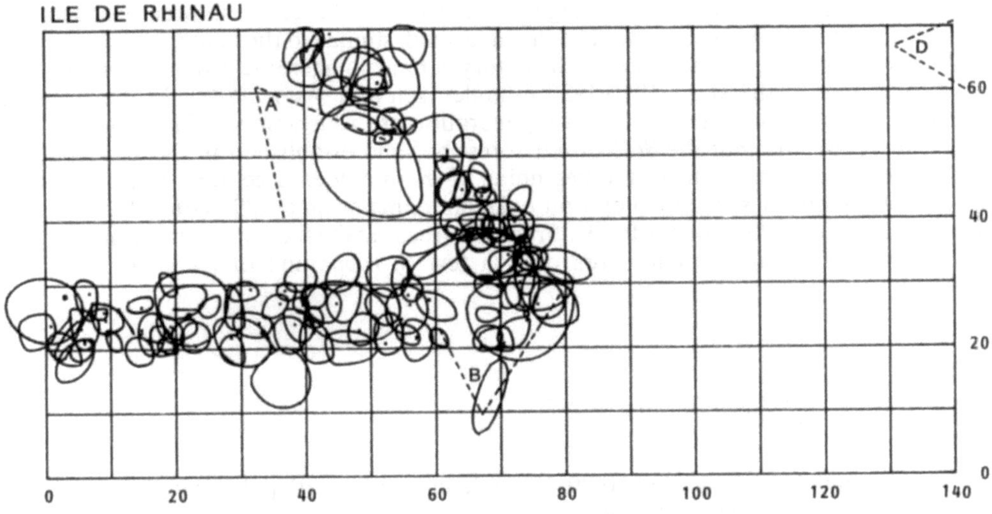
ILE DE RHINAU
A
B
D
0
20
40
60
80
100
120
140
20
40
60

A
B
C
D

Fig. 8.3. Computer-plotted ground plan drawing of the area sampled for tree rings in the core area of Ile de Rhinau (France)(previous page). **A** The stream gulley allong the transect area inwhich the tree rings have been sampled. **B** View on the transect with the old oak at the left. Positions of the photographs (June 1984) are indicated in the ground plan. **C** Big canopy gap at the end of the core area overgrown by *Clematis vitalba* and climbing in the trees. **D** Under the weight of the *Clematis vitalba* blanket young trees have been downed and are reitterating

No traces of uprooting have been found except for the two oaks that were uprooted in 1982. If there were any mounds and pits of possible former uprootings, these have probably been washed away by the frequent floods. With the tree ring data four retrospective views at 15 years' time intervals have been plotted in Figure 8.4. Only six trees, one *F. excelsior* and five *Q. robur* trees, are older than 80 years. It is very likely that at this time the core area still contained remainders of the pioneer *Salix-Alnus* forest. Because of the low durability of their wood, no remnants of tree bodies of those species can be found any more at present. In 1926 two *U. laevis* trees and one *Alnus glutinosa* x *incana* became established. Furthermore, one *Populus nigra*, two *P. alba* and two *P. tremula* trees became established. The spread of *Populus* spp. probably was favoured by their ability of clonal rejuvenation (Koop 1987). Four *F. excelsior* trees established themselves.

At the time of the first picture in 1936 a huge regeneration wave of *F. excelsior* is starting. At this moment probably the *Salix-Alnus* forest had wholly collapsed, while in the mean time the river dynamics declined due to the construction of the Grand Canal d'Alsace. In the core area only few *Salix alba* trees survived, possibly by means of intermediate vegetative reproduction. Because of their fast juvenile growth, the *P. alba* and *P. nigra* trees have overgrown the *F. excelsior* trees and the oldest *Alnus glutinosa*. Both *P. tremula* trees have a lead over the *F. excelsior* trees.

In the picture of 1951, a group of seven *Alnus glutinosa* x *incana* trees appears and still new *F. excelsior* becomes established. A *Prunus fruticans* and *Prunus padus* established themselves together with these *Alnus* spp. In the picture of 1966 two of the younger *F. excelsior* trees at 30 m along the baseline reached the dimensions of the much older *F. excelsior* (at 60 m). The *F. excelsior* trees underneath a canopy of *Q. robur* are lagging behind the other ashes. Two other *P. padus* became established.

The picture of 1981 shows that the *Populus tremula* trees remained behind in growth and have the same height as the *F. excelsior* trees that formerly became established underneath them. The *P. alba* and *P. nigra* trees have become the highest trees of the transect. Only a few *F. excelsior* trees reached a dominant position in the canopy. The oaks are the oldest trees. They gradually grew on and kept their dominant role in the canopy from the beginning, although *Populus* spp. and *F. excelsior* grew much faster. It is obvious that the leaning *P. alba* and *P. nigra* trees at about 40 m along the baseline grew away from the big *Q. robur* crown just above the transect. The *P. alba* at 100 m surmounted the neighbouring *Q. robur* before this tree had spread its crown over the *P. alba* stem.

8.1.5 Mosaic Development During Regeneration

In conclusion, the historically documented *Alnus-Salix* forest soon after establishment became invaded by *Quercus robur* and *Ulmus* spp. A second regeneration

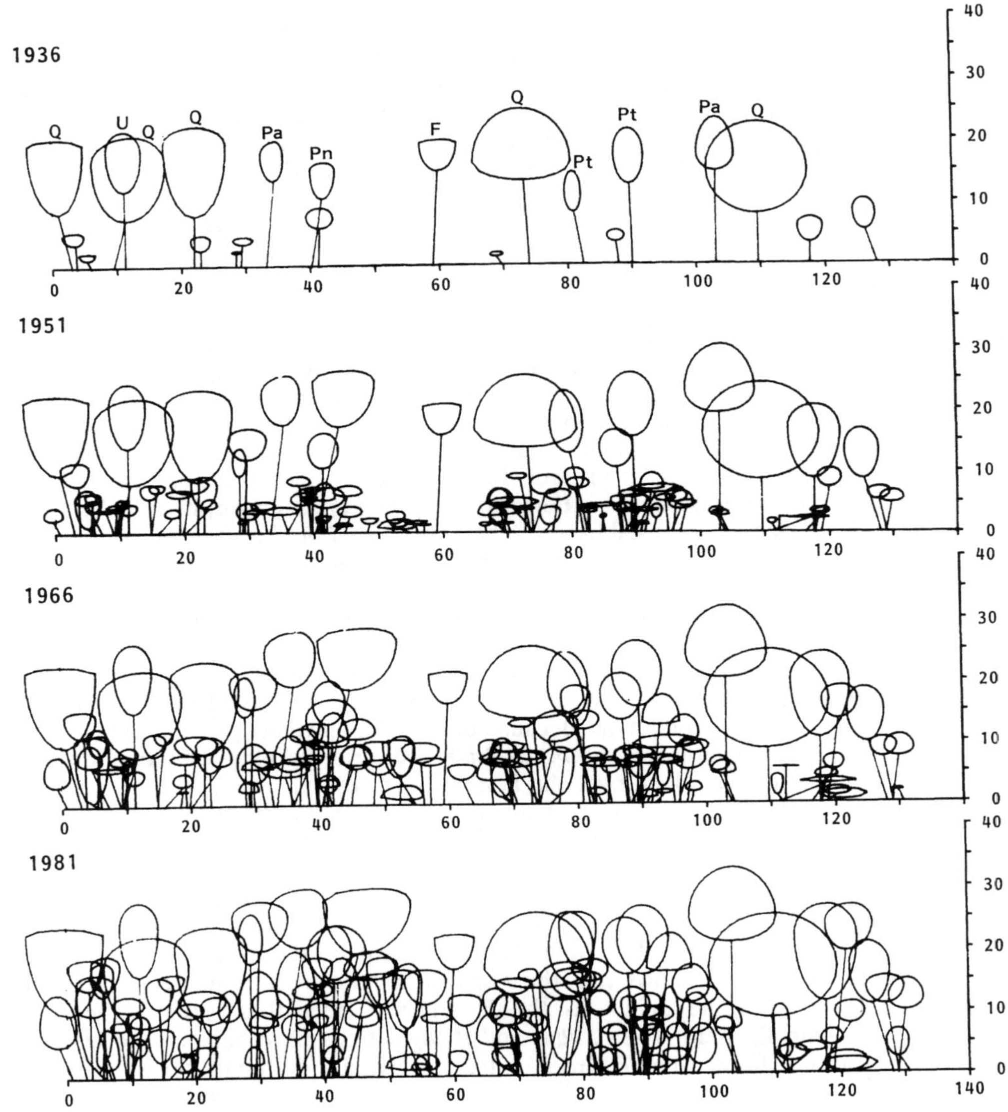

Fig. 8.4. Reconstructed structural forest development using the program GROWRING (see Chap. 6.5). Woody transect area in Core area of Ile de Rhinau. Four pictures at 15 years intervals have been cut out of the cartoon film that covers 80 years. *Q Quercus robur*, *U Ulmus* spp., *Pa Populus alba*, *Pn Populus nigra*, *Pt Populus tremula*, *F* and not indicated *Fraxinus excelsior*

wave of *Q. robur* and *Ulmus* spp. and the first establishment of *Populus* spp. trees around 1880 indicate the starting collapse of the pioneer forest. In the 1930's and 1940's, *F. excelsior* invaded the forest. Only a few older *F. excelsior* trees were found in the core area. At that time, the *Salix-Alnus* forest had probably collapsed, while in the meantime the river dynamics declined due to the construction of the Grand Canal d'Alsace. Although the forest structure suggests an uneven-aged structure, three distinct tree generations can be recognized from the tree ring analysis. Different tree species with their own growth rate and maximum life-span resulted in a complex forest structure. In comparison with the core areas of Fontainebleau, it is striking that, due to more than one tree species and various but higher tree growth rates, generation changes take less time and the homogeneous regeneration unit of *Salix* and *Alnus* fragmented much quicker.

8.2 Successional Forest Dynamics in Unexploited Alder-Birch Swamp Forest in the Otterskooi (The Netherlands)

The forest dynamics of the Otterskooi will be reconstructed in a different way, i.e. without using such tree ring data as were available for the forests dealt with in the previous paragraphs. The reconstruction will be based upon interpretation of historical data and present structural features such as spatial distribution patterns of tree species, patterns in architectural compartments and tree diameter distribution. Because of a serial succession process the relation with ground vegetation will be stressed. The core area of the Otterskooi was first charted in 1982. Every 3 years the herbaceous transect plots have been reassessed and major changes in forest structure have been recorded.

8.2.1 Historical Analysis

The Otterskooi is an ancient duck-decoy area in the Dutch nature reserve "de Wieden" in the northwestern part of the province Overijssel (6°.02'EL, 52°.36'NW) owned by De Vereniging tot Behoud van Natuurmonumenten ("Association for the Conservation of Natural Monuments"). The heart of the decoy area consists of a number of excavated pools. The ducks were lured to the pools and moved through funnels connected to the pools leading to small cages to be entrapped. In a silence zone within a radius of 1130 meters around the pools hunting is prohibited by law. For isolation and silence the direct surroundings of the decoy pools were often afforested. In the Middle ages, the region of the "Wieden" was an ombrotrophic peat-moor. After the sixteenth and seventeenth century the peat was cut. First the cutting took place above groundwater level. Canals were dug for transport and drainage. The canal called "Thyssen gracht", that delimits the Otterskooi on the north side, can be first traced on a map of 1650 (Dijkstra et al. 1986). Later, when the peat surface declined, peat was dredged from below groundwater level, dried and also used for fuel. According to the soil map at scale 1:200,000 (Stichting voor Bodemkartering 1965) the Otterskooi is mapped as a complex of more or less open ditches and sometimes man-made peat soils. This complex originates from the former way of peat cutting from below groundwater level, in mainly mesotrophic and oligotrophic peat lands. The spoil and peat was dredged from long straight ditches and dried ashore on narrow embankments of the uncut peat between the ditches. The ditches, when not too deep and wide, filled up with gyttja, originating from peat erosion and from aquatic and shore vegetation. After some time the ditches became overgrown by a floating swamp or fen vegetation forming a dense network of roots and rhizomes. This floating swamp layer became thicker and more solid. Then it became

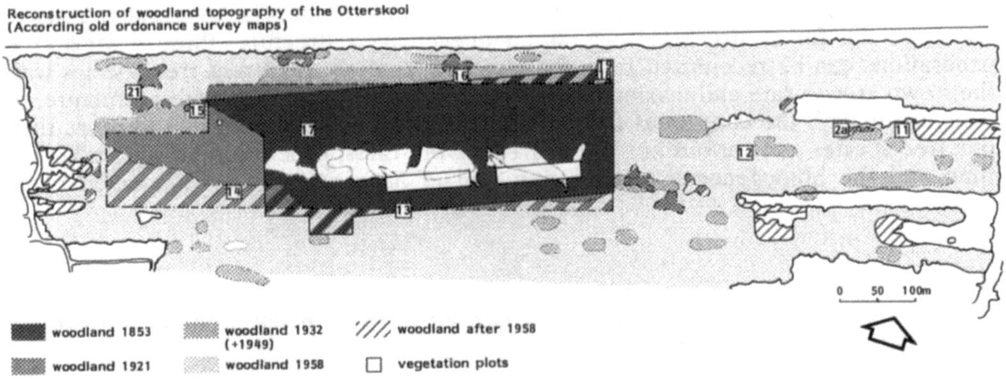

Fig. 8.5. Reconstructed topography of the forest reserve the Otterskooi (Overijssel, The Netherlands)

passable and could even bear trees. The complex of embankments of untouched peat, the so-called "ribben" ("ribs") and formerly more or less open ditches between them form a recognizable feature of the soil pattern of the Otterskooi. A topographical reconstruction of the decoy forest according to the land registration map of 1826 and the ordnance surveys of 1853, 1921, 1932, 1949, 1958 is given in Figure 8.5.

First the floating swamp or fen vegetation described above was used for the production of reed for thatching purposes. Progressive woodland development could start after termination of mowing management, since the beginning of the 20th century (Wiegers 1985). Because of an earlier stop of the mowing management in the direct surroundings of the decoy, between 1826 and 1853, the decoy forest of the Otterskooi is at least 130 years old and thus belongs to the oldest fen-woodlands in The Netherlands. Vlieger (1937) described the vegetation of the decoy forest and its surrounding grasslands and made two vegetation relevés. Two photographs made of the vegetation plots and his description of the decoy forest indicate that the forest had not been actively managed for a long time. In 1939 the decoy and its direct surroundings received the status of nature reserve (Thijsse 1939). Ever since, no active management has taken place. The topography of 1949 is virtually similar to that of 1932, although this may be due to an incomplete revision. The map of 1958 shows a considerable forest increase and decreased surface of the two most western decoy pools. Except for some small later changes the forest area has remained the same since then.

8.2.2 Vegetation Pattern and Forest History

Collation of the forest history mapping with the vegetation map (Fig. 3.9) teaches that the *Thelypteris* unit, the *Calamagrostris* unit as well as the *Carex* unit and the complexes they form, are found in the younger woodlands. The *Calamagrostris* variant of the *Dryopteris* type is found in the young woodlands at the northern edge of the forest originating from a *Sphagnion* vegetation (see Westhoff and den Held 1969), as is

Fig. 8.6. A The *Dryopteris* unit in the core area of the Otterskooi (The Netherlands, June 1984). **B** The *Impatiens* unit in the core area of the Otterskooi (The Netherlands, June 1984)

documented by Vlieger (1937). The *Impatiens* type, the typical variant and the *Phragmites* variant of the *Dryopteris* type are confined to the oldest forest part that has existed at least 100 years. The occurrence of the *Impatiens* type (Fig. 8.6) can be attributed to the contact between the sites and open slowly flowing water which formerly was led through the decoy pools. The water intake came through the ditch east of the decoy pools. The ditch in the western edge of the present core area served as the outlet to the Thijssen gracht. Only recently has this connection been dammed to prevent tourists entering the forest by boat. The direct contact with the water of the canals outside the reserve has its influence on the eutrophy of the water just below the floating soil layer, as has been proved by Van Wirdum (1979) for open fenlands not far from the Otterskooi. Moreover, the eutrophy of the water is probably raised by regular duck feeding and repeated spoil dredging. The typical *Dryopteris* unit (Fig. 8.6) is found around the decoy pools and in the northern fringe area of the forest. A decomposing top soil, well drained because of the vicinity of open water, or piled up from spoil dredged from the ditches or decoy pools, determines the range of this unit.

8.2.3 Forest Architecture and Forest History

A relation between increasing age of forest stands going from the core to the edge of the forest complex and the patterns of forest structure and species composition of canopy trees might give an indication for the timing of its development. Forest structure mapping (Fig. 8.7) as described in Chapter 3 shows larger coherent surfaces of tree phase (height over 20 m) around the decoy pools. Away from the pools, the tree phase occurs in small scattered spots in a mosaic with thicket phase, poletree phase and gap phases.

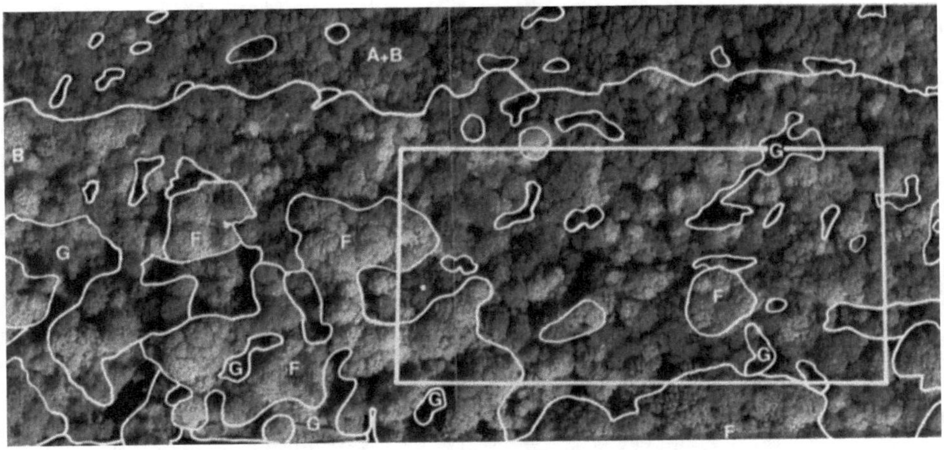

Fig. 8.7. Forest phase mapping. The following phases are distinguished: gap phase (*G*), pole tree and thicket phase (*B*), tree phase (*F*) and a homogeneous pole tree phase of young, spontaneous forests (A+B). The location of the one-hectare core area is indicated

West of the core area a divergent mosaic with a high percentage gap phase with poletree phase is mapped. This part of forest was used by cormorants (*Phalacrocorax carbo*) in a breeding colony during the period 1944-1948 (Veldkamp 1985). The colony reached its maximum size of about 2000 breeding pairs in the years 1945 and 1946. Part of the *Alnus glutinosa* canopy trees died and the gaps were not filled up again. An extensive poletree phase has been mapped in the first-generation spontaneous stands at the periphery of the forest. The limit between this homogeneous poletree phase and the heterogeneous mosaic of poletree-, tree- and gap phases, coincides with the limit between the oldest decoy forest and the forest enlargement since 1958.

8.2.4 Tree Species Distribution and Forest History

A maximum likelihood classification (Schowengerdt 1983) based on the green, red and infrared digitized colour bands from true colour and black and white infrared, 60 x 60 mm photographic negatives, has been carried out to distinguish tree species (see Chap. 6.4). A set of training pixels, collected from the reconstructed aerial view (program CANOPIX) of the core area is used. Figure 8.8 shows the end image of tree species distribution in an area around the core area. The first-generation forests, in the extreme northern part of the analyzed area, shows the highest amount of *Betula*. In the older forests in the southern part their share decreases. The portion of *Alnus* slightly decreases southward. In the south, near the decoy pools the amount of *Quercus* and *Fraxinus* increases, gradually forming coherent patches. In the northern part only a few isolated small *Fraxinus* and *Quercus* crowns were located. So the highest proportions of the last two species are found in the oldest forest.

8.2.5 Architectural Compartments and Tree Distribution

In the core area, forest history and potential development can be analyzed using the present data of forest structure, such as diameter distribution diagrams, in relation

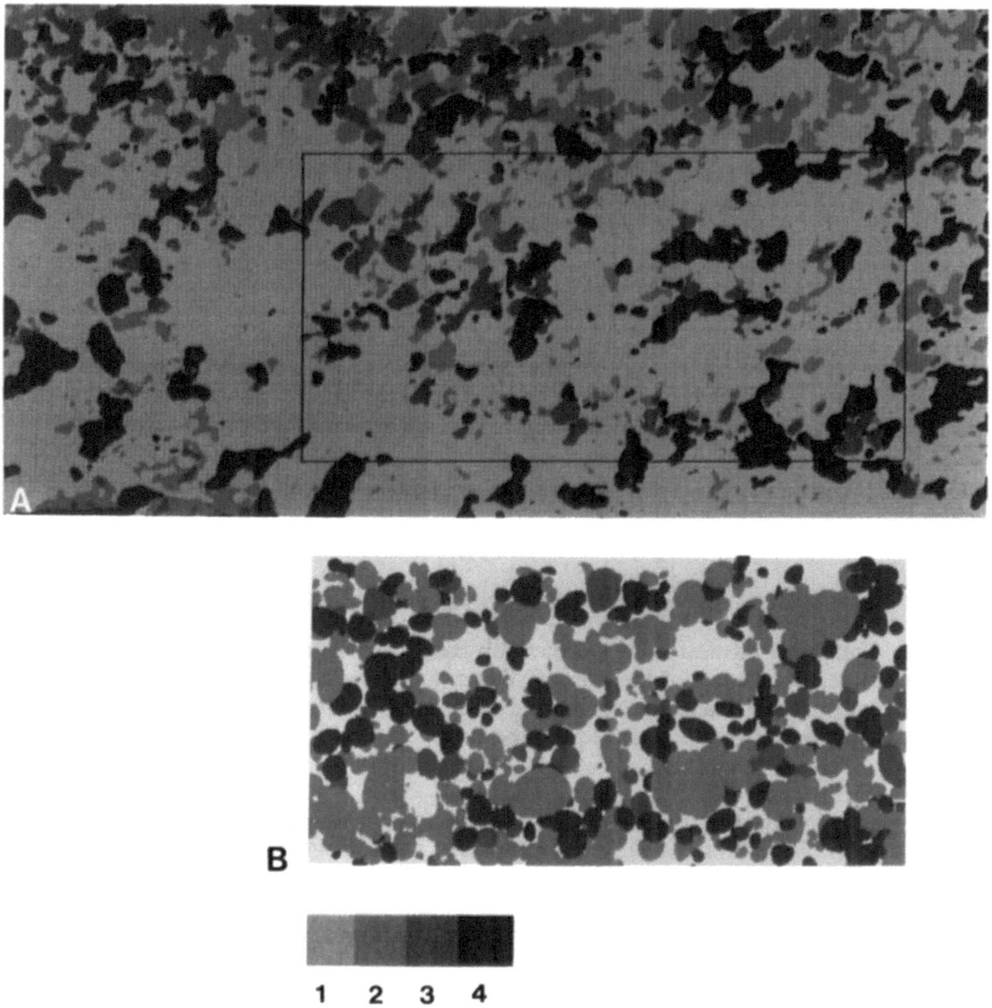

Fig. 8.8. A Tree species map generated by digital processing of aerial photographs. *1* = *Alnus glutinosa*, *2* = *Betula pendula*, *3* = *Quercus robur* and *Fraxinus excelsior* sun exposed and in the shade, *4* = *Sorbus aucuparia* and *Salix multinervis* and canopy gaps. **B** Simulated aerial view from the core area as indicated in **A**. For the species the same grey values are used

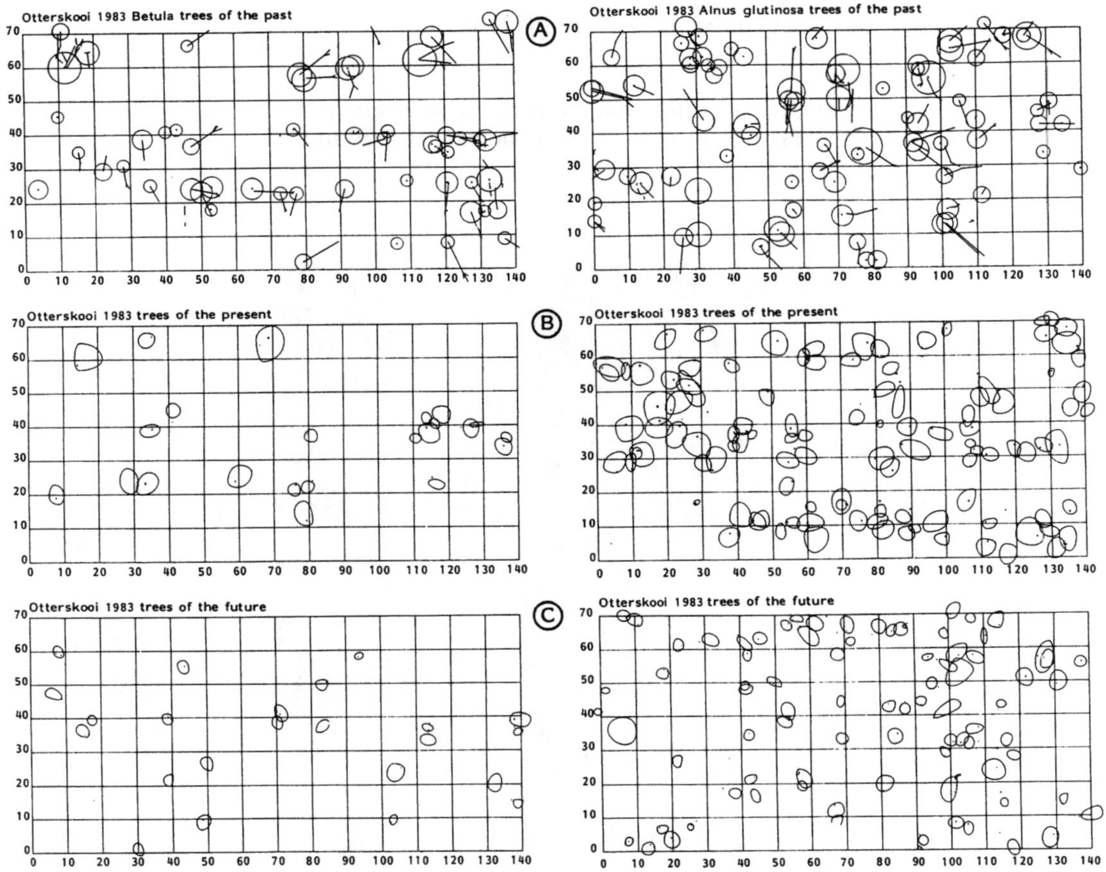

Fig. 8.9. Computer-plotted crown projection maps for *Betula* spp. and *Alnus glutinosa* of: **A** Reconstructed trees of the past. **B** Trees of the present. **C** Potential trees. Reconstruction took place by attributing a circular crown projection to all standing stumps proportional to the diameter of the bole

to the vegetation pattern in the core area. The distribution patterns of all tree species are separately plotted for two different architectural compartments (see Chap. 3): i.e. the set of potential trees and the set of trees of the present. A third graph is made of reconstructed crown projections of the trees of the past. In order to do so, a crown is assigned to dead trees and uprooted and broken dead trees with a diameter proportional to the stem diameter (Figs. 8.9, 8.10, 8.11).

The distribution patterns are worked out by plotting the tree densities and relative portions of potential trees, trees of the present and trees of the past per species against the vegetation types (Figs. 8.12, 8.13). Vegetation types are therefore arranged in a sequence as derived from the first axis of a DECORANA ordination of relevés of the vegetation types. The first axis of this ordination can be interpreted as stretched between open water with wet and thin peat layers towards thick superficially dessicated peat layers. The distribution patterns of the tree species per architectural code and diameter distribution will be analyzed below.

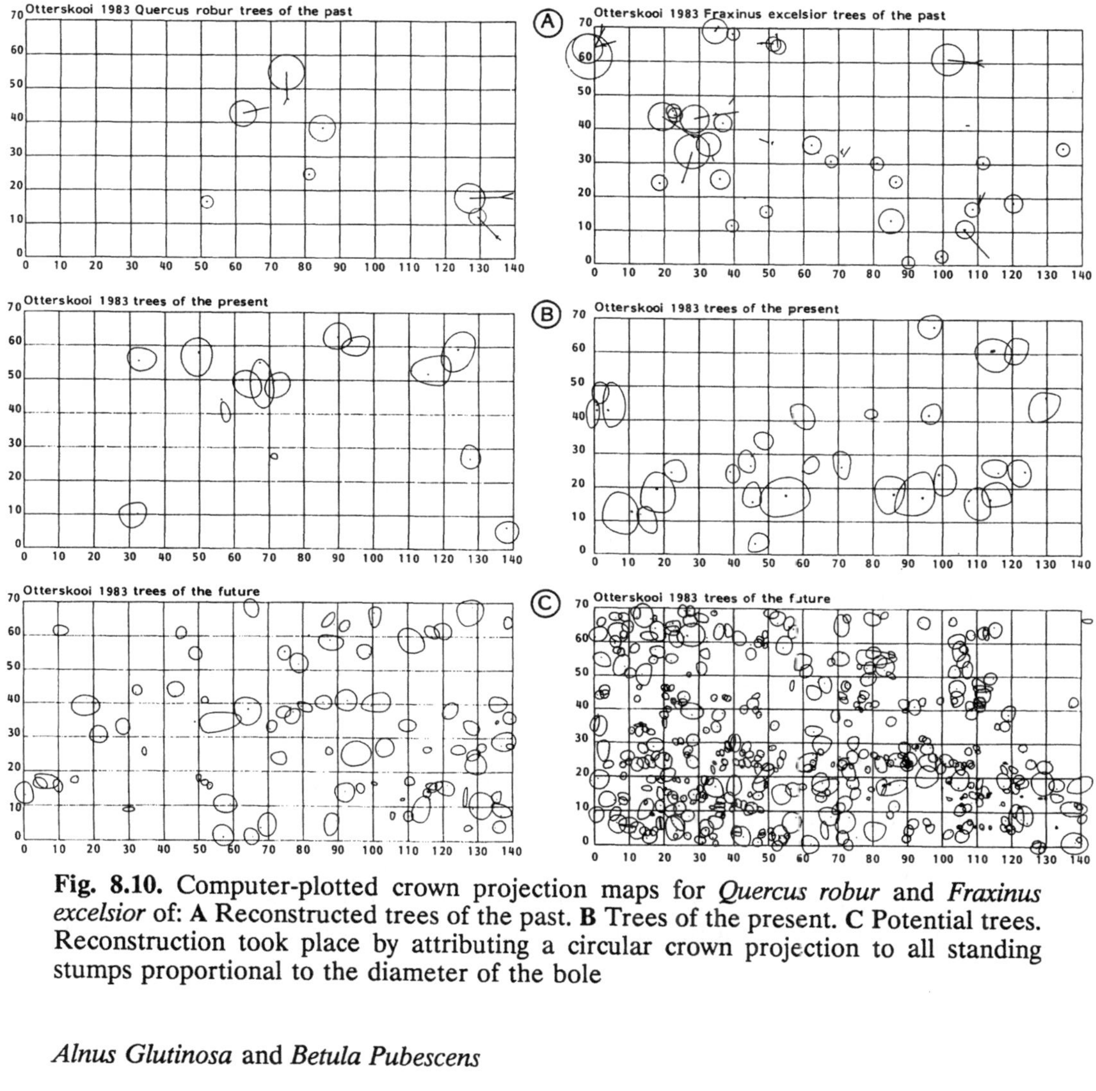

Fig. 8.10. Computer-plotted crown projection maps for *Quercus robur* and *Fraxinus excelsior* of: **A** Reconstructed trees of the past. **B** Trees of the present. **C** Potential trees. Reconstruction took place by attributing a circular crown projection to all standing stumps proportional to the diameter of the bole

Alnus Glutinosa and *Betula Pubescens*

The diameter distributions of *Betula pubescens* and *Alnus glutinosa* are very similar (Fig. 8.14), although the absolute numbers of *B. pubescens* are four times lower than of *A. glutinosa*. (Fig. 8.12). A maximum tree number is found in the diameter classes between 20 and 40 cm. Lower diameter classes show lower numbers. This means that the present dominant role of both species in the tree canopy cannot be maintained in the future. Vegetative reproduction through suckers of *A. glutinosa* might offer some temporary compensation, because the advantage of an established root system gives them a better chance to succeed in reaching the canopy than newly established saplings (Koop 1987). Figure 8.13 shows that the relative portion of trees of the past is about the same in all vegetation types, except for the ditches and the *Iris* type. In the ditches and the *Iris* type, where trees of the present of *A. glutinosa* occur only at lowest densities, densities of the potential trees are highest (Fig. 8.12). Potential trees of *A. glutinosa* have their highest relative shares in the ditches and the *Iris* type (Fig. 8.13). Lowest densities of

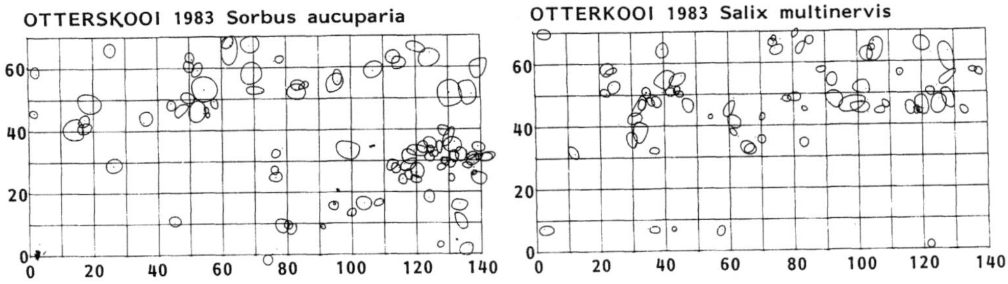

Fig. 8.11. Computer-plotted crown projection maps for *Sorbus aucuparia* and *Salix multinervis*

potential trees occur in the *Calamagrostris* variant and typical variant of the *Dryopteris* type (Fig. 8.12). The relative portion of trees of the present, however, is highest in the *Calamagrostris*, the typical and *Phragmites* variant of *Dryopteris* type (Fig. 8.13). In the *Sorbus* and *Phragmites* variant of the *Dryopteris* type higher tree densities of trees of the present and potential trees coincide. Trees of the present of *B. pubescens* show a strong affinity to the *Sorbus* type (Figs. 8.12 and 8.13). Tree densities decrease in the *Phragmites* and typical variant of the *Dryopteris* type respectively. Potential trees of *B. pubescens* are relatively scarce and show a distribution pattern similar to the trees of the present. Trees of the past show a similar pattern but their densities are higher than those of the present and potential trees. While trees of the past were present in all types except in the ditches, trees of the present occur only in the typical and *Phragmites* variants of *Dryopteris* type and in the *Sorbus* type (Fig. 8.13). As a potential tree it plays a subordinate role only in all vegetation types. In conclusion, the role of *A. glutinosa* and *B. pubescens* declines. Trees of the present of *A. glutinosa* still indicate the dominant role of the tree in the present situation. Potential trees of *A. glutinosa*, however, play a role of some importance only on the younger, thin peat soils of the ditches and the *Iris* type. These only cover small surfaces. The higher relative proportions of potential trees of *A. glutinosa* and *B. pubescens* in the *Phragmites* and *Sorbus* type can possibly be explained by a lower competition factor of *F. excelsior* at these sites.

Quercus robur and *Fraxinus excelsior*

Diameter distributions of *Fraxinus excelsior* and *Quercus robur* (Fig. 8.14) differ fundamentally from those of *B. pubescens* and *A. glutinosa*. Both species have highest tree numbers in the lower diameter classes. The tree numbers in higher diameter classes decrease. This distribution warrants a lasting role of both species in the future. Because of the higher numbers of *F. excelsior* this tree will probably be the first to take over the dominant role of *A. glutinosa*.

Potential trees of *Quercus robur* are present in all vegetation types except on the youngest peat soils in the ditches and show a distribution gradient. High densities are found in the *Sorbus* type and low densities in the *Iris* type (Figs. 8.12 and 8.13). The trees of the present and trees of the past are only found in the typical and *Phragmites* variant of the *Dryopteris* type and in the *Sorbus* type. The shares of the trees of the past, trees of the present and potential trees in the vegetation types mentioned are low (Fig. 8.13).

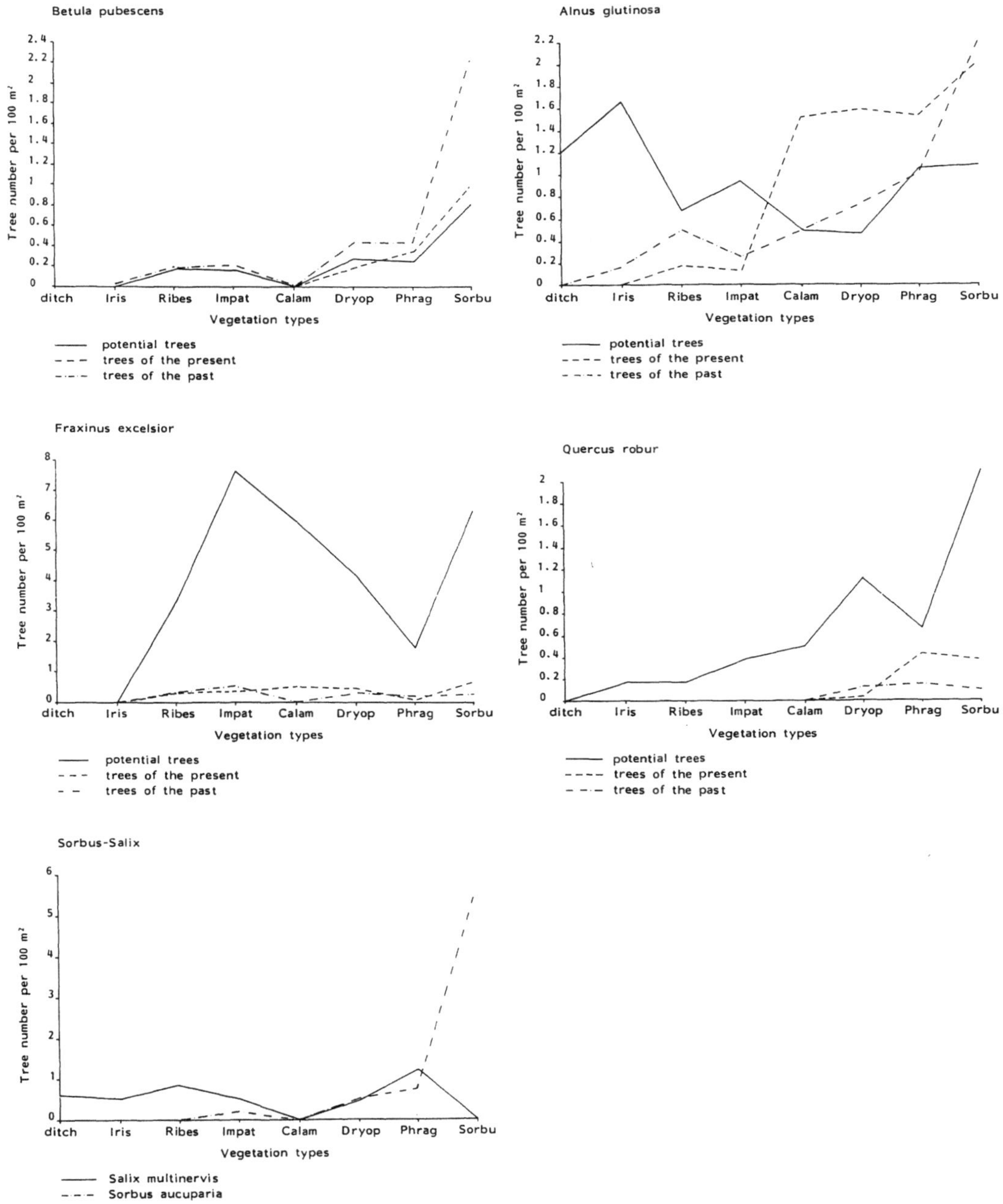

Fig 8.12. Tree numbers per 100 m^2 of *Alnus glutinosa*, *Betula pubescens*, *Fraxinus excelsior* and *Quercus robur* per growth potential against vegetation type. For *Salix multinervis* and *Sorbus aucuparia* no difference was made between trees with different growth potential

In conclusion, *Q. robur* first established itself on the oldest peat soils, that first became dry. Later establishment of the species in vegetation types with younger soils took place as these places had become dryer.

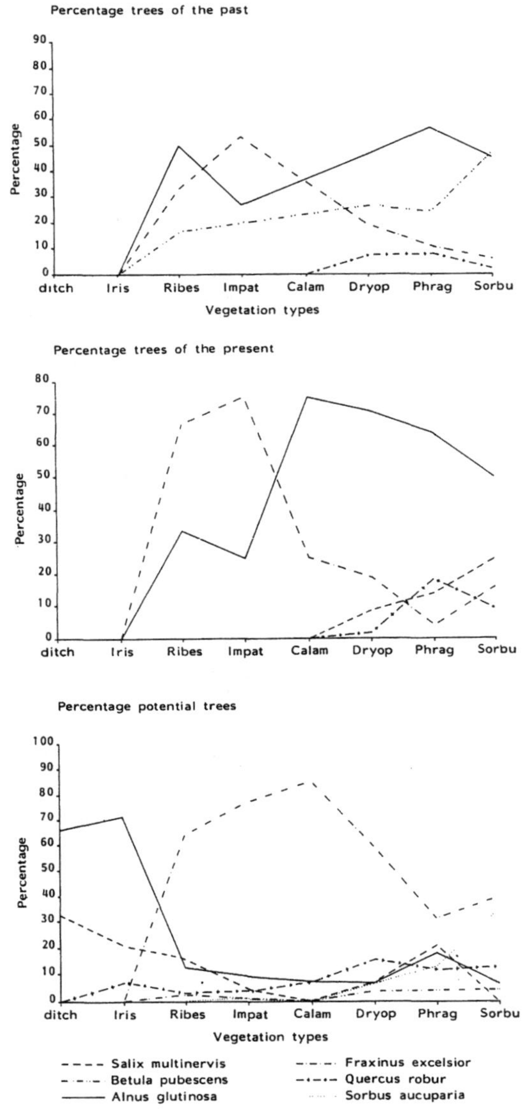

Fig. 8.13. Proportion of trees of the past, trees of the present and potential trees of *Alnus glutinosa*, *Betula pubescens*, *Fraxinus excelsior* and *Quercus robur* against vegetation type

Fraxinus excelsior has an inverse distribution pattern with regard to *Sorbus aucuparia*, *B. pubescens* and *Q. robur* (Figs. 8.12 and 8.13). In both potential trees and trees of the present, a clear distribution gradient from most trees in the *Impatiens* type, less in the typical and *Calamagrostris* variant and least in the *Phragmites* variant of the *Dryopteris* type can be distinguished. The occurrence of trees of the present and potential trees in the *Sorbus* type is caused by their preference for the edges of the ditches in contact to the open water, that have been charted as *Sorbus* type. In fact, these narrow bands show a steep gradient of different microsites. *F. excelsior* is the most frequent potential tree in all vegetation types. Exept in the *Iris* type, in its whole range of occurrence *F. excelsior* has the highest proportion in potential trees (Fig. 8.13). In trees of the present it plays a dominant role only in the *Ribes* and *Impatiens* type, while in trees of the past it dominates only in the *Impatiens* type. In conclusion, *F. excelsior* invades the dryer types from the *Impatiens* type. On the youngest peat soils of the *Iris* type and the ditches and on the other side further away from the open water in the *Phragmites* variant and in the *Sorbus* type the role of *F. excelsior* as a potential tree declines.

Sorbus aucuparia

Sorbus aucuparia is found mostly in the *Phragmites* variant and in the direct border zone of this variant to the typical variant of the *Dryopteris* type in the north-eastern part of the core area (Fig. 8.11) and of course most frequently in the *Sorbus* type itself. In lower diameter classes *S. aucuparia* has a distribution of diameters similar to *F. excelsior* and *Q. robur* (Fig. 8.14). Many thin trees warrant a stable future population.

Salix multinervis

Salix multinervis occurs most frequently in relatively open soil depressions along the ditches in the *Phragmites* variant of the *Dryopteris* type and the *Impatiens* type and in the ditches themselves (Fig. 8.11). Together with *A. glutinosa*, it is the only species that grows in the ditches. *S. multinervis* has a diameter distribution with highest tree numbers in the classes around 10 cm (Fig. 8.14). Both thinner and thicker trees are less frequent. In regard to this distribution it might seem that the future population of the species is not provided for. However, because of layering and the partly creeping behaviour of the trunks, trunks of a certain maximum diameter never rise above the 1.30-m level at which the Dbh is measured. Nevertheless, the species permanently rejuvenates vegetatively.

Although the 5-year period (1982-1987) during which this core area is under study using the information system is still far too short for general conclusions on forest dynamics, the changes observed in this period confirm the developmental trend derived from the forest structure analyses. Thirteen *Alnus glutinosa* trees, seven *Betula pubescens* trees, one *Fraxinus excelsior* and one *Sorbus aucuparia* fell down during this period. In the monitoring sequence, the next survey of the whole core area, including height and diameter measurements, will take place in 1992.

8.2.6 Potential Natural Vegetation

With the help of the architectural and historical analysis of the forest vegetation, it is possible to predict a future development in terms of potential natural vegetation (PNV) (see Chap. 3). In terms of PNV the vegetation types can be divided into three units. The *Thelypteris* unit and the *Carex* unit and their complex, confined to the younger woodlands (see Fig. 3.9), as well as the ditches and the *Iris* type, on the youngest thin peat layers in the core area, can be considered as an *Alnetum*. In these units the highest portions of *B. pubescens* and *A. glutinosa* have been detected on the aerial photograph

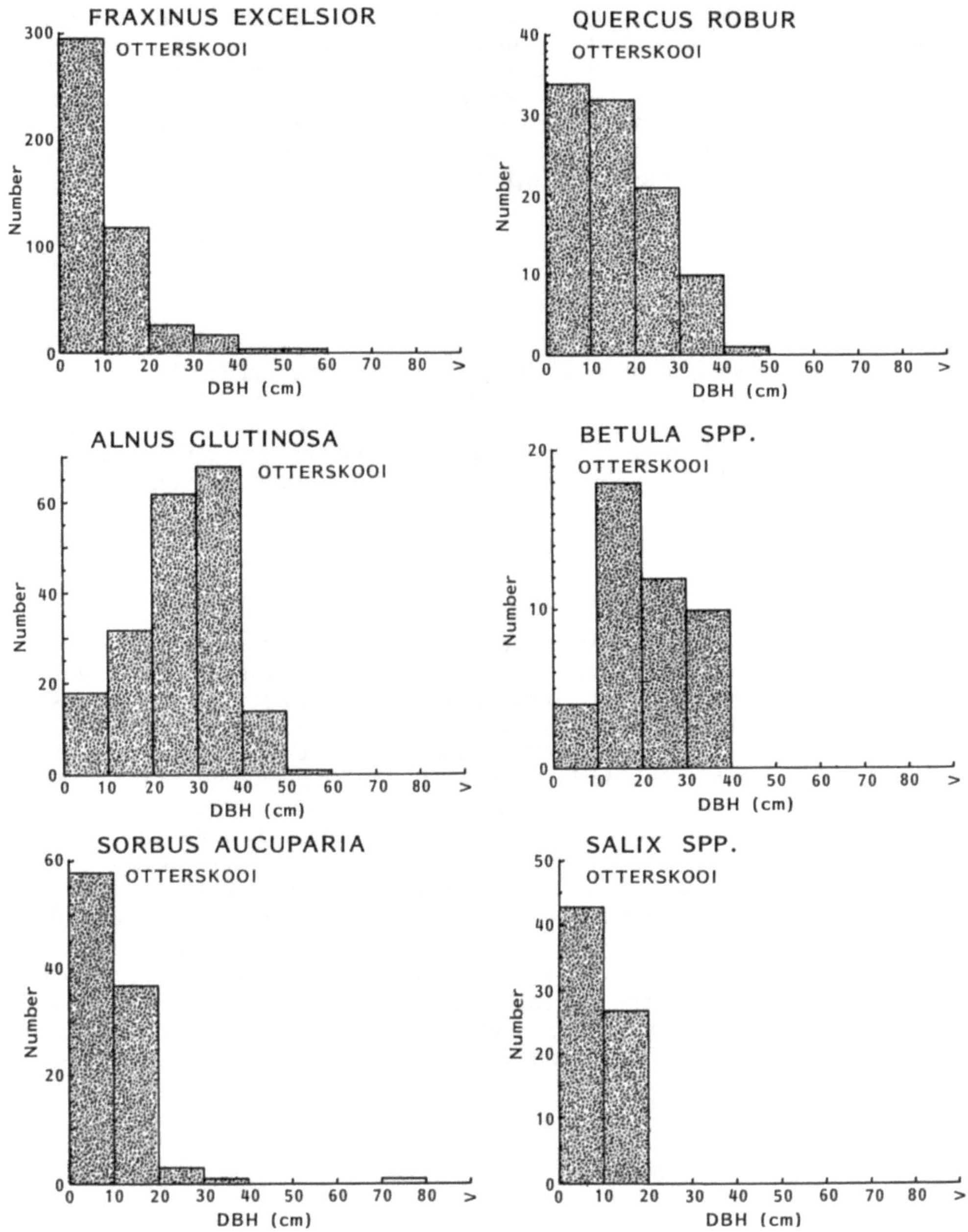

Fig. 8.14. Diameter distribution diagram of *Fraxinus excelsior*, *Quercus robur*, *Alnus glutinosa*, *Betula* spp., *Sorbus aucuparia* and *Salix multinervis*

(Fig. 8.8). In the core area the trees of the past are scarce (Fig. 8.12), which proves the portions of *B. pubescens* and *A. glutinosa* have been detected on the aerial photograph first-tree generation nature of the stands. At present there is no indication of the end of the dominant role for *A. glutinosa*. Compared to phyto-sociological literature, the local vegetation types described above can be best referred to as a *Thelypteris* and a *Carex* variant of the *Carici elongatae-Alnetum* (Van der Werf 1989), the first one being closely related to the *Thelypterido-Alnetum* (Klika 1940).

The *Impatiens* unit, confined to the area with flowing water in and below the floating peat soil, shows the highest portions of *F. excelsior* and *Q. robur* in the upper canopy on the aerial photograph (Fig. 8.8). In the core area in the *Impatiens* and *Ribes* type, the dominant or co-dominant role of *A. glutinosa* in the trees of the past has been taken over by *F. excelsior* in the trees of the present (Fig. 8.13). In the future the set of potential trees will reinforce the dominance of *F. excelsior*. The state of the present forest gives evidence for a development from an *Alnetum* to a *Fraxinetum*. Phyto-sociologically, the local *Ribes* and *Impatiens* types can be regarded as a transition between the *Carici elongatae-Alnetum* and the *Pruno-Fraxinetum* (Van der Werf 1989).

The *Dryopteris*- and *Phragmites* unit, confined to the oldest decoy forest away from the contact zone to flowing water. In species composition of the upper tree canopy, they are intermediate between the two units first discussed (Fig. 8.8). In the core area in the *Dryopteris* type and especially its *Phragmites* variant, *A. glutinosa* is the dominant tree of the present. In the set of trees of the past besides *A. glutinosa*, *B. pubescens* played a co-dominant role. In the set of potential trees, however, *B. pubescens*, *Q. robur* and *F. excelsior* have higher shares than *A. glutinosa* (Fig. 8.13). *A. glutinosa* seems to be retreating and in the future this forest will probably be a mixture of *Q. robur*, *B. pubescens*, *A. glutinosa* and *F. excelsior*. According to Van der Werf (1989), phyto-sociologically the *Dryopteris* type can be best referred to as a subassociation *dryopteritosum* of the *Carici elongatae-Alnetum*, in transition to the *Lysimachio-Quercetum* (Pasarge 1968). This vegetation type shows great similarity to the *Dryopterido-Alnetum* distinguished by Klika (1940).

The *Sorbus* unit grows on floating peat soil with a rather thick litter layer, that is most superficially desiccated. In the core area, in comparison to the *Dryopteris* type, *F. excelsior* has no role of importance in the *Sorbus* type. Its occurrence is restricted to the edges of the unit in transition zones to the ditches. The local *Sorbus* type with a mixture of *B. pubescens*, *A. glutinosa*, *Q. robur* and *S. aucuparia* as potential trees and trees of the present (Fig. 8.13) phyto-sociologically can be considered as a *Lysimachio-Quercetum* (Van der Werf 1989).

The *Calamagrostris* unit is found in the young woodlands at the northern edge of the forest and originates from a *Sphagnion* vegetation (Vlieger 1937). It shows clear phyto-sociological relationship to the *Alno-Betuletum* (Van der Werf 1988). In the core area this variant, however, is only represented as tiny fragments and no conclusions can be drawn from the tree stand.

The forest development deduced from the present forest architecture and species composition has been collated with succession theories in literature (Fig. 8.15). According to Smittenberg (1969, 1976) and Wiegers (1985), a succession line from the *Carici elongatae-Alnetum thelypteridetosum* to the subassociation *caricetosum acutiformis* and finally to the "dry variant" can be expected. The subassociation *betuletosum pubescentis* is expected to develop to the dry variant as well. The authors do not mention a separate subassociation with *Dryopteris*, but the *Dryopteris* type might be considered as part of the "dry variant". The "dry variant", which syntaxonomically was not named by Smittenberg and Wiegers, probably can be identified as the *Lysimachio-Quercetum* (Pasarge 1968; Van der Werf 1989) which is a quite recently described syntaxon.

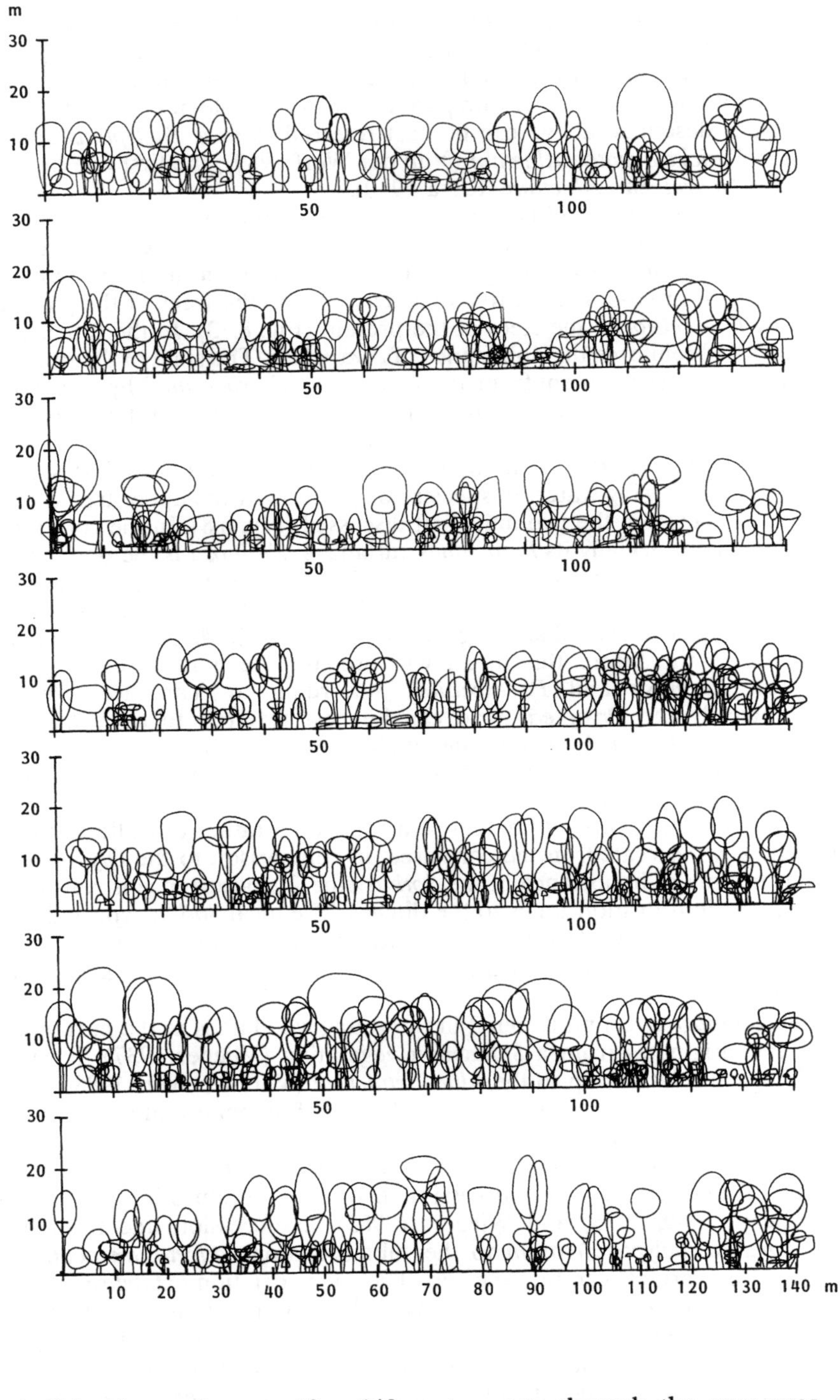

Fig. 8.16. Side views of seven 10 x 140 m transects through the core area of the Otterskooi (Overijssel, The Netherlands)

Carici elongatae Alnetum (Smittenberg 1969)
thelypteridetosum
(*Thelypteris* type)
|
v
Carici elongatae Alnetum
caricetosum acutiformis
(Smittenberg 1969)
(*Carex* variant of the *Dryopteris* type)
|
v

-->

Sphagnion
|
v
Carici elongatae Alnetum
betuletosum pubescentis
(Smittenberg 1969)
(*Calamagrostris* variant of the *Dryopteris* type)
|
v

"dry variant"
(*Dryopteris* type and *Sorbus* type)
Lysimachio-Quercetum
(Pasarge 1968; Van der Werf 1989)

Fig. 8.15. Schematic representation of vegetational succession in fen-woodlands according to Wiegers (1985) as far as relevant to the vegetation types described for the Otterskooi (in brackets)

Together, the patches of the PNV's of *Carici elongatae-Alnetum*, *Pruno-Fraxinetum* and the *Lysimachio-Quercetum* form a fine-grained site mosaic which is highly determined by the pattern of ditches and "ribs" and the distance to flowing water. The hierarchy of the sylvatic mosaic as a subsystem of the site mosaic has been interchanged here (see Chap. 2). The forest architecture has different potentials at short distance, which results in a heterogeneous and diverse forest structure and species composition (Fig. 8.16). Because succession has not yet ended in a steady state, it seems likely that the patches of *Carici elongatae-Alnetum* will develop into a *Pruno-Fraxinetum* or a *Lysimachio-Quercetum* in the future.

Chapter 9

Monitoring of Forest Dynamics

Tens of years or even centuries of observation are necessary to detect forest dynamics as expressed by architecture and tree species composition (Chap. 7). However, fast initial changes in structure and species composition can be exemplified in young, spontaneous willow stands in Zuidelijk Flevoland, a polder reclaimed in 1968 (The Netherlands). To detect changes in the herb layer composition long sequences of relevés of the 2 x 2 m herbaceous transect area, quadrats would be needed. The oldest relevés now in the information system were first analyzed in 1982 as the project started. Because the Neuenburger Urwald was subject to an earlier study in 1978 and was included in the information system later, data on the Neuenburger Urwald have been used for an interim analysis. The results of this earlier interdisciplinary study (soil science, silviculture and phyto-sociology) in 1978 (Koop 1981) will be briefly summarized before discussing the results of the time series.

9.1 Stand Dynamics in a Spontaneous Willow Forest in Zuidelijk Flevoland, a Polder Reclaimed in 1968 (The Netherlands)

In 1968, the polder Zuidelijk Flevoland was reclaimed. In the construction of the dikes, mats of woven willow rods were used which formed suckers. As soon as the land stood clear of water, seeds produced by these suckers germinated on the bare moist soil. Hoogesteger (1983) used aerial photographs to describe the initial stages of forest development over the area now in the forest reserve of the Wilgenreservaat. In the following years after the initial germination on the bare soil, no other trees became established. Recently, some shrubs of *Sambucus nigra* and *Crataegus monogyna* established themselves.

The Wilgenreservaat is situated along the Knardijk, the north-eastern enclosing dike of the polder Zuidelijk Flevoland. A 0.7 core area was first measured in 1983. The maximum tree height in the reserve is less than 20 m. Therefore the width of the 140-m-long core area was restricted to 50 m, resulting in an surface area of 0.7 ha (see Chap. 2)(Figs. 9.1, 9.2, 9.3). The changes owing to tree death in the core area were reassessed in 1986. These data can be used to show changes in species composition and forest structure.

During the period 1983-1986 mortality was highest among *Salix triandra*, which is a shrubby species (Table 9.2) Among the trees that died we find a high percentage of bendt trees, i.e. trees with their tree base beyond their crown projection, especially under *S. triandra* and *S. alba* x *fragilis*. *Salix alba* showed the lowest percentage of bended trees in 1983, only 14% (Table 9.1). Fifty percent of the trees of this species that died during the period 1983-1986 were bent. Among *S. triandra* the dead trees had a bigger crown surface than those who survived. The bending of the low shrubby species increases the

Fig. 9.1. Photographs of the core area Wilgenreservaat (Province Flevoland, The Netherlands, June 1986). The position the photographs were taken is indicated in Figure 9.3. **A** Dead shrubby *Salix triandra* under an open canopy of *S. alba*. **B** Canopy gap caused by the death of *Salix triandra* shrubs

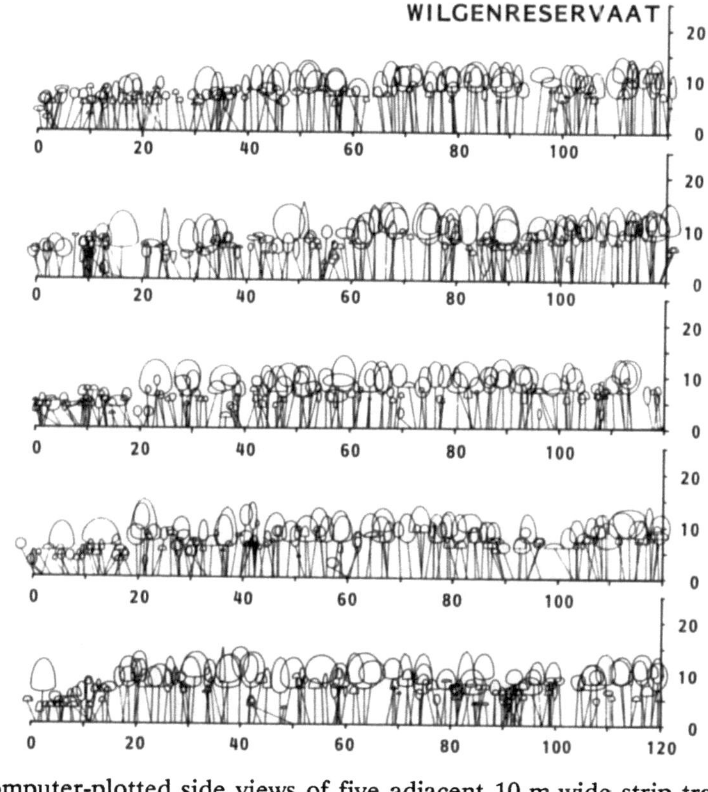

Fig. 9.2. Computer-plotted side views of five adjacent 10-m-wide strip transects in the core area Wilgenreservaat (Province Flevoland, The Netherlands). For reasons of clarity only 120-m length of the transects have been plotted

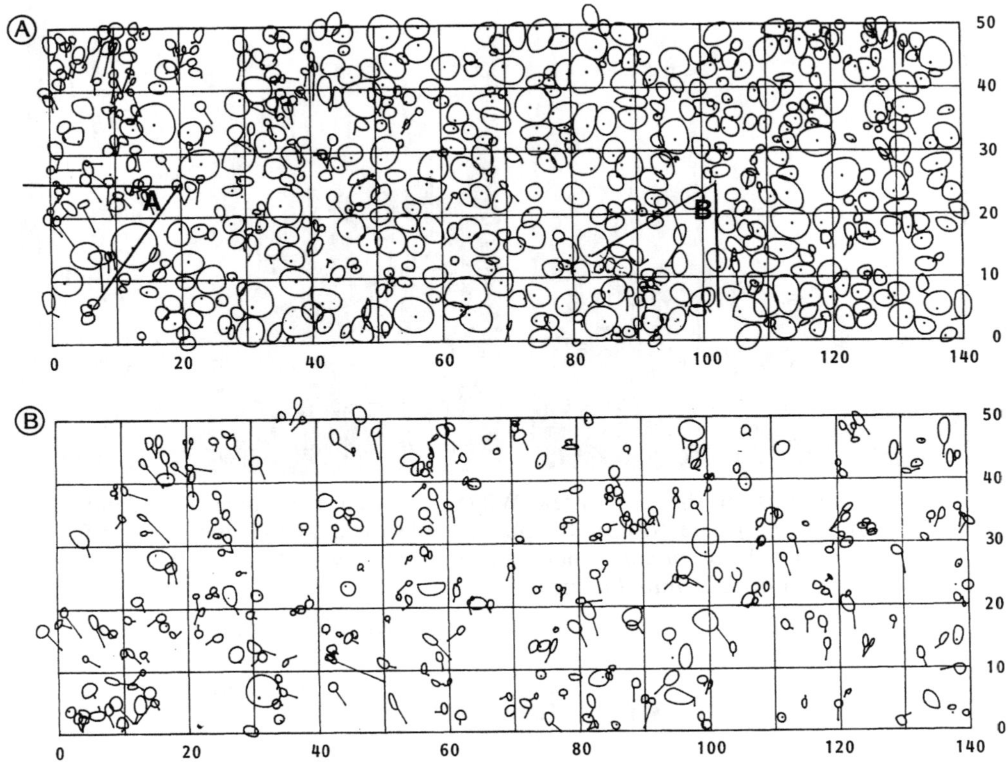

Fig. 9.3. Computer-plotted plans of the core area Wilgenreservaat (Province Flevoland, The Netherlands). The position of the photographs in Fig. 9.1 is indicated. **A** Situation 1983. **B** Trees that died in the period 1983-1986

crown surface. In *S. alba* and *S. alba* x *fragilis* the trees that died had a smaller crown surface than the surviving trees. The dead trees of the latter species were higher than those of *S. triandra* and had flat crowns. Bending did not increase the crown surface. Only one shrub of *Sambucus nigra* became established in the willow forest already before 1983. Seed sources of other tree species (*Populus* spp., *Fraxinus excelsior*, *Quercus robur*, *Acer pseudoplatanus* and others) have become available in forest plantations at a distance of about 500 meters. In 1983 all 50 adjacent quadrats are covered by 100 % *Urtica dioica* only. In 1986 in some plots the cover of *U. dioica* had declined but was not replaced by any other species nor tree seedling. Whether other tree species will be able to colonize the forest before the willows die, and what species these will be, are the main questions to be answered by future monitoring. It is quite possible that the willow forest collapses before other trees have become established. An early collapse might be enhanced by the bacterial disease *Erwinia salicis*.

Table 9.1. Total tree number (*N*), Basal area (*G*), Crown area (*CA*) and the percentage of bent trees per hectare of the core area of the Wilgenreservaat (Province Flevoland, The Netherlands) in 1983

Trees	N	G(m^2)	CA (m^2)	%Bent
Salix triandra	1005	6.14	1462	61
Salix alba x *fragilis*	77	1.54	567	20
Salix alba	456	8.71	3340	14

Table 9.2. Tree number (*N*), Basal area (*G*), Crown area (*CA*) and the percentage of bent trees per hectare of all trees that died in the period 1983-1986 in the core area of the Wilgenreservaat (Province Flevoland, The Netherlands)

Trees	N	G(m^2)	CA(m^2)	%Bent
Salix triandra	397	2.36	671	82
Salix alba x *fragilis*	7	0.04	9	80
Salix alba	48	0.56	127	50

In another reserve with similar spontaneous willow forests, the Lepelaarsplassen, transects were drawn by De Coo (1981) and repeated by Burger in 1988 (unpublished). The changes in those transects which were incorporated in the present information system in 1988, are in line with those in the core area of the Wilgenreservaat.

9.2 Herb-Layer Dynamics in Unexploited Beech Forests in the Neuenburger Urwald (Federal Republic of Germany)

The Neuenburger Urwald, a formerly grazed forest between Wilhelmshaven and Oldenburg in Niedersachsen (F.R.G.) has been maintained as a forest reserve since 1870.

Humus-iron-podsol soils (Mückenhausen 1977a, 1977b) in sandy deposits of the Saale glacial period carry a beech-oak forest (*Fago-Quercetum*). Gley brown earths in loamy sands carry a wood millet-beech forest (*Milio-Fagetum*). Gley-pseudogley brown earths in loam, with a stagnating water table on a subsoil, consisting of silty to clayish basin deposits of the Elster glacial period, carry an oak-hornbeam forest (*Stellario-Carpinetum typicum*). The influence of uprooting upon soil conditions and herbaceous species composition has been analyzed. Forest architecture was clearly correlated with soil conditions and forest type. In order of decreasing homogeneity of structure the forest types were ranked, i.e. beech-oak, wood millet-beech and oak-hornbeam forest respectively. Similarly, the size of the widest canopy gaps decreases from more than two times tree height, to one-and-a-half to two times tree height, to one to one-and-a-half times tree height, in the same order.

Because of the heterogeneous structure of the forests, the floristic composition changes within one forest type. In poletree and thicket phases the number of species is generally lowest. Under small canopy gaps the number of species increases, while in the biggest gaps the number decreases again, because low herbs of the tree phase are superseded by the increasing competition of brambles, honeysuckle or ferns. Because of decreased root competition and absence of interception of precipitation, under canopy gaps species of moist forest types occur in dryer forest types. Beech and holly are the two most important tree species that will determine forest structure in the oak-beech and the wood millet-beech forest in future. Holly maintains itself in the understorey of the beech forests and grows quickly in canopy gaps, whatever their size, unless it is destroyed by falling crowns. In the wood millet-beech forest, holly disappears under larger canopy gaps because of competition from brambles and honeysuckle, or because mice girdle the

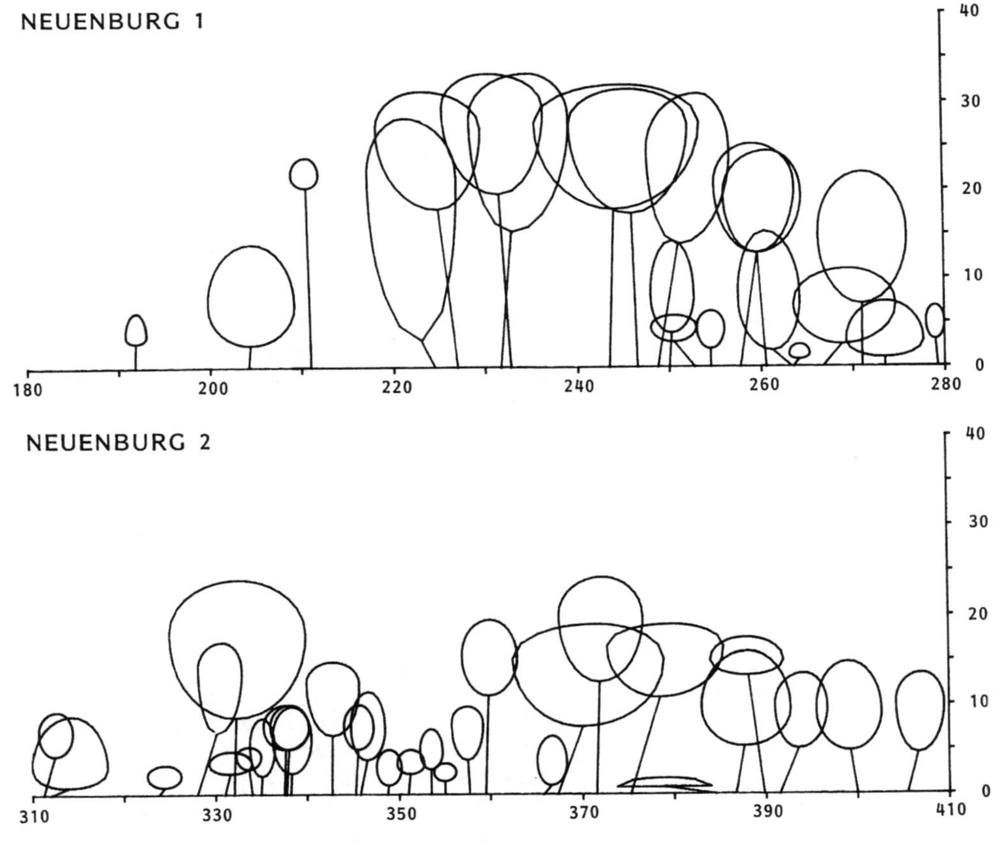

Fig. 9.4. Computer-plotted transect of the transect intervals **A** 180-280 m and **B** 310-410 m in the Neuenburger Urwald from Koop (1981), described as *Fago-Quercetum* and *Stellario-Carpinetum*, respectively

Table 9.3. Frequency percentage of species in the four clusters distinguished in the three relevés of 1978, 1981 and 1985 of the *Fago-Quercetum* transect interval 180-280 m in the Neuenburger Urwald. *1*=*Rubus*, *2*=*Dryopteris*, *3*=*Ilex* and *4*=*Milium* cluster

Cluster	1	2	3	4
Number of quadrats	32	29	68	21
* *Rubus fruticosus*	91	45	3	5
Athyrium filix femina	47	17	3	0
Rubus idaeus	56	52	0	0
* *Sorbus aucuparia*	25	31	7	0
* *Dryopteris dilatata*	16	52	0	0
Dryopteris carthusiana	6	62	6	5
Lamium galeobdolon	3	24	0	0
Fagus sylvatica	0	7	1	43
* *Milium effusum*	0	7	1	57
Anemone nemorosa	0	0	1	57
Oxalis acetosella	9	62	9	67
Carpinus betulus	44	55	37	14
* *Hedera helix*	75	93	91	71
Ilex aquifolium	16	62	87	57
Agrostris stolonifera	6	0	0	0
Scutellaria galericulata	0	3	0	0
Carex remota	0	3	0	0
Galeopsis tetrahit	0	3	0	0
Acer pseudoplatanus	0	3	0	0
Deschampsia flexuosa	0	0	13	0
Quercus robur	0	0	1	0
Lonicera periclymenum	0	3	3	10

stems in the grassy environment of the clearing. In the oak-beech forest young beeches develop best under the widest canopy gaps. In the wood millet-beech forest they develop best in the relatively small gaps because competition from brambles is too strong in the larger gaps. Based on observations, a forest development cycle is reconstructed. Vegetational, structural minimum areas, in which all representative stages and phases of forest development continue to be present by self-regulation of the forest, were deduced from forest mosaic textures of the different forest types. These are certainly not minimum areas for a number of fauna elements.

Sets of three sequential relevés of quadrats of similar site type were subject to both DECORANA ordination and TWINSPAN clustering (Hill 1978a;b). Within the vegetation types *Fago-Quercetum*, *Milio-Fagetum* and *Stellario-Carpinetum*, which were described as different sylvatic mosaics by Koop (1981), transect intervals of 100 m were selected (Fig. 9.4). Data on vegetation composition of only higher plants of the quadrats were available for the years 1978, 1981 and 1985. Per site type, the set of all sequential relevés of the herbaceous transect area quadrats were clustered with the program TWINSPAN and cluster transition during the period 1978-1985 is depicted in cluster transition bars (Figs. 9.6 and 9.8). The species composition of the clusters is represented in Tables 9.3 and 9.4. Species groups have been distinguished and will be referred to by the name of one species that is marked in the table with an asterisk (*). The scores of the relevés against the first two DECORANA axis are plotted (Figs. 9.6 and 9.8). Although the clusters in the diagrams most often overlap, a delineation for relevés belonging to the same cluster is drawn. Transition vectors that connect different year relevés of the same quadrat were drawn to indicate shifts of quadrats through the diagram. For reason of clarity of the figure, however, not all transition vectors could be drawn, but only those that show shifts from one to another cluster.

9.2.1 Herb-Layer Dynamics in a *Fago-Quercetum*

From aerial photographs, before the big storms of 1972 and 1973, the transect interval 180-280 m proved to be a quite homogeneous beech forest (cf. Koop 1981). In the storms, part of it (180-220 m) was thrown over. Data on species composition before the storms are not available. Because of similar soil conditions and similar forest structure before the storm, proved by aerial photographs and forest reconstruction by graphically erecting the thrown beeches (Koop 1981), it seems likely that the species composition did not differ too much from the actual species-poor composition in the unaffected part of the transect interval (220-280 m). TWINSPAN clustering of all relevés for the years 1978, 1981 and 1985 of the quadrats resulted in four clusters (Fig. 9.6, Table 9.3).

The first cluster (referred to as Rubus cluster) has high frequency and coverage of the *Rubus fruticosus* group and the *Sorbus aucuparia* group. Except for the *Hedera helix* group, all other species have a low frequency (Table 9.1). This cluster was found in the wind-thrown part of the transect in 1978 (Fig. 9.5). It appeared as a continuous blanket of lianes, except for some quadrats underneath the *Carpinus betulus* tree that had survived the storm. In 1981 and 1985 the extent of quadrats belonging to the *Rubus* cluster decreased gradually. In 1985 it was only found in the centre of the widest remaining canopy gap. The species of the *R. fruticosus* group are light-demanding, while the presence of *Athyrium filix femina* reflects the higher soil moisture in a gap (Koop 1981).

The second cluster (referred to as *Dryopteris* cluster) shows a remarkably lower frequency of the *Rubus fruticosus* group. Together the *Rubus* cluster and the *Dryopteris* cluster are characterized by the *Sorbus aucuparia* group. High frequency of the *Dryopteris dilatata* group and *Oxalis acetosella* is typical for the *Dryopteris* cluster. The *Dryopteris* cluster has spread since 1978 and succeeds the *Rubus* cluster (Fig. 9.5). In the closed beech forest, isolated patches appeared and disappeared underneath small canopy gaps.

The third cluster (referred to as *Ilex* cluster) is characterized by the highest cover of *Ilex aquifolium* in a low shrub layer. Except for *Deschampsia flexuosa* and species of the *Hedera helix* group all other species have a low presence. The *Ilex* cluster is found under a closed canopy of the beech forest in the transect interval 220-266 m. At the edge of the clearing in 1985 a transition from the *Rubus* cluster towards the *Ilex* cluster could be observed. This transition was due to closing of the canopy of *Ilex* shrubs, which were released from a beech canopy during the storm. Before the storms the *Rubus* cluster in the open part of the transect interval will probably have been preceded by the *Ilex* cluster, as is now found in the second unaffected part of the transect. A frequent transition *Rubus-Dryopteris* occurred. In a few cases there was a transition from the *Rubus* to the *Dryopteris* back to the *Ilex* cluster. Therefore the sequence *Ilex*-, *Rubus*-, *Dryopteris*- back to the *Ilex* cluster can be regarded as the "most complete cycle" after the disturbance caused by the 1972-1973 storms.

The fourth cluster (referred to as *Milium* cluster) consists of the last seven quadrats at the end of the transect interval (266-280 m). They represent a transition to the *Milio-Fagetum*. It is characterized by high coverage of *Oxalis acetosella* and the *Milium effusum* group. During the years 1978-1985 its species composition hardly changed. Only in two quadrats did a transition to the *Ilex* cluster take place, probably caused by dense shade.

In Figure 9.6b the scores of all relevés on the first and second DECORANA ordination axis are plotted. The sequence of the *Rubus*, the *Dryopteris* and the *Ilex* cluster is clearly found along the first axis. This allows the first axis to be interpreted as canopy cover. Simultaneously the nitrogen indication according to Ellenberg (1974) decreases as well for the specific groups (6.2-5.7-5.0), which can be explained by nitrification of the litter layer. Coverage in the *Ilex* cluster is high in the tree canopy. Differences in coverage between the *Rubus* and the *Dryopteris* cluster are caused by closing regeneration of *Ilex aquifolium*, *Sorbus aucuparia* and *Fagus sylvatica*. It is striking that part of the *I. aquifolium* as well as the *F. sylvatica* treelets regenerated vegetatively (Koop 1987). Figure 9.6 shows the increasing values of the first DECORANA axis in time. The sequence on the second axis corresponds with the transition to another sylvatic unit.

Fig. 9.5. Photographs of the herbaceous transect interval 180-190 in Neuenburger Urwald (Niedersachsen, F.R.G.). **A** In June 1977 the ground vegetation was dominated by Rubus fruticosus *(Rubus* cluster). **B** In June 1985 the same area carried a *Dryopteris dilatata*-dominated vegetation (*Dryopteris* cluster)

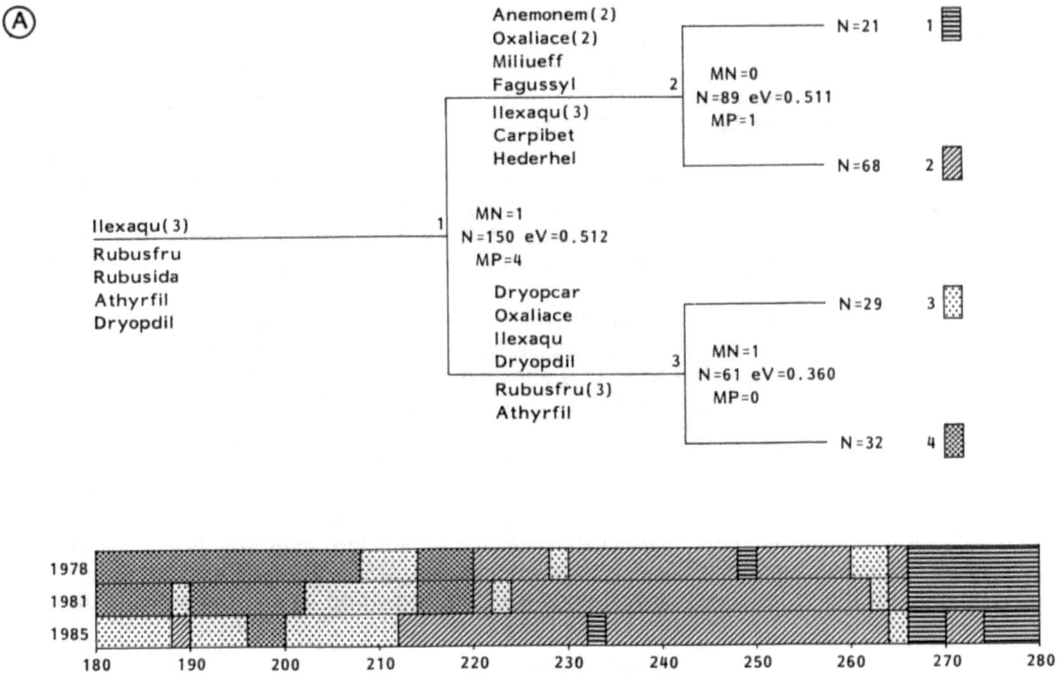
A
Anemonem(2)
Oxaliace(2)
Miliueff
Fagussyl
Ilexaqu(3)
Carpibet
Hederhel
N=21
1
MN=0
N=89 eV=0.511
MP=1
2
N=68
2
Ilexaqu(3)
Rubusfru
Rubusida
Athyrfil
Dryopdil
1
MN=1
N=150 eV=0.512
MP=4
Dryopcar
Oxaliace
Ilexaqu
Dryopdil
Rubusfru(3)
Athyrfil
3
N=29
3
MN=1
N=61 eV=0.360
MP=0
N=32
4
1978
1981
1985
180
190
200
210
220
230
240
250
260
270
280

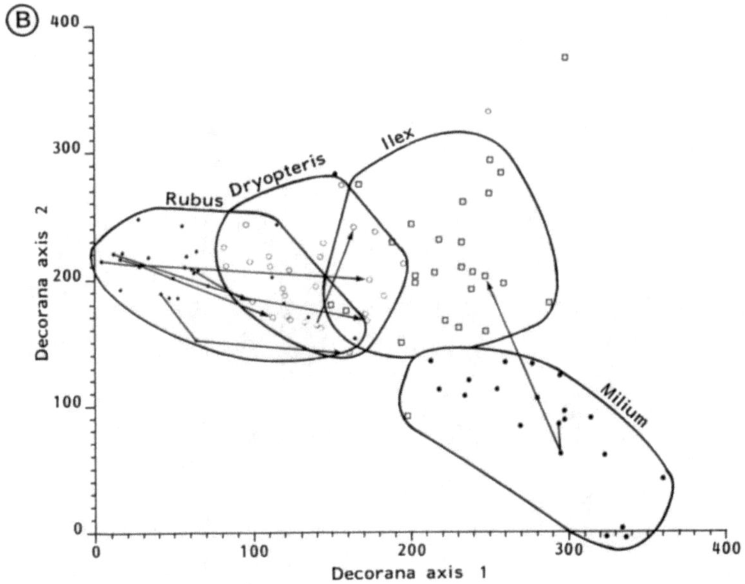
B
Decorana axis 2
Decorana axis 1
Rubus
Dryopteris
Ilex
Milium
0
100
200
300
400

Fig. 9.6. **A** TWINSPAN dendrogram and cluster transition bar of the set of sequential relevés of the herbaceous transect area of the *Fago-Quercetum* 1978-1981-1985. *1=Milium, 2=Ilex, 3=Dryopteris* and *4=Rubus* cluster. **B** Plotted scores of the relevés against the first two DECORANA axis. Relevés belonging to the same TWINSPAN cluster are delineated. Transition vectors that connect different year relevés of the same quadrat are drawn to indicate shifts of quadrats through the diagram. For reasons of clarity of the figure, however, not all transition vectors could be drawn, but only those that show shifts that are cluster transitions

9.2.2 Herb-Layer Dynamics in a *Stellario-Carpinetum*

The transect interval 310-410 m is a heterogeneous mixed oak-hornbeam forest (*Stellario-Carpinetum*). During the period of observation 1978-1985 a canopy gap in a poletree phase in the interval 312-326 closed by a spread of neighbouring trees crowns. In the transect interval at two places a crown of a tree from beyond the transect limits fell into the transect. In 1984 an old oak fell down and created a new canopy gap. Four clusters were distinguished (Fig. 9.8 Table 9.4).

The first cluster (referred to as *Oreopteris* cluster) is characterized by high frequency and cover of ferns, mainly *Oreopteris limbosperma*. Lower species such as those of the *Oxalis* group are very scarce except for *Anemone nemorosa*, that flowers before the ferns close their canopy and then disappears for the rest of the year. With the *Stellaria* cluster this cluster has the *A. nemorosa* group in common. *Phegopteris connectilis* most frequently appears in this cluster. The species grows exclusively on the mounds of uprootings (Koop 1981). Together with the cluster as a whole the species, however, fades away as the site becomes overcast. As the canopy gap closes during the years 1981 and 1985, relevés of this cluster disappear. On the most shaded spots they move to the *Hedera* cluster and in the edges of the poletree phase to the *Stellaria* cluster.

A second cluster (referred to as *Stellaria* cluster) is widely spread throughout the transect interval, mostly underneath oak trees (Fig. 9.5). It is characterized by the highest number of species and the highest cover of the *A. nemorosa* group. Minor shifts from the *Stellaria*- to the *Hedera* cluster occurred during the period of observation.

The third cluster (referred to as *Hedera* cluster) is the cluster poorest in species. Only species of the *Oxalis acetosella* and the *Dryopteris dilatata* group occur. All light-demanding species such as those of the *Rubus fruticosus* group are scarce. The relevés belonging to this cluster occur underneath a closed canopy mostly of *Carpinus betulus* or of a shrub layer of *Ilex aquifolium* or *Fagus sylvatica*. Relevés of all other clusters can change towards this *Hedera* cluster when they suffer from heavy overcast (Fig. 9.5).

The fourth cluster (referred to as *Lonicera* cluster) is characterized by a high frequency of the *Rubus fruticosus* group just as the *Stellaria* cluster, but it has much lower frequencies of the *A. nemorosa* and the *O. acetosella* groups. Opposite to the *Oreopteris* cluster with a dense canopy of big ferns, here the smaller ferns *Dryopteris carthusiana* and *D. dilatata* play an important role. Although very rare, the species *Polypodium vulgare*, *Deschampsia flexuosa* and *Blechnum spicant* indicate the rather acid character of the relevés. Quadrats belonging to this cluster are found in the areas of rotting woody debris of more or less collapsed crowns of uprooted trees.

In the DECORANA graph (Fig. 9.8b) the first axis can be interpreted as canopy cover. Low values coincide with the shaded *Hedera* cluster and high values with the open *Oreopteris* cluster. In this respect the *Lonicera* cluster is partly covered by collapsing tree crowns. The explanation of the second axis might be acidity. The most extreme values

Fig. 9.7. Photograph of the herbaceous transect in the *Stellario-Carpinetum* in Neuenburger Urwald (Niedersachsen F.R.G.) June 1985. **A** *Hedera* cluster in transect interval 310-320 m. **B** *Stellaria* cluster in transect interval 330-350 m

Table 9.4. Presence percentage of species in the four clusters distinguished in the three relevés of 1978, 1981 and 1985 of the *Stellario-Carpinetum* transect interval 310-410 m in the Neuenburger Urwald. *1*=*Oreopteris*, *2*=*Stellaria*, *3*=*Hedera* and *4*=*Lonicera* cluster.

Cluster	1	2	3	4
Number of quadrats:	11	72	34	31
Oreopteris limbosperma	100	1	0	0
Phegopteris connectilus	27	19	0	0
Lamium galeobdolon	36	60	6	3
Deschampsia cespitosa	73	68	20	29
Athyrium filix femina	36	58	26	13
* *Anemone nemorosa*	91	80	18	3
Milium effusum	64	92	12	16
* *Oxalis acetosella*	0	94	47	42
Stellaria holostea	18	69	12	10
Ilex aquifolium	0	18	32	10
Carpinus betulus	0	75	94	32
Hedera helix	64	35	50	13
* *Dryopteris dilatata*	45	50	23	61
Dryopteris carthusiana	0	29	35	55
Lonicera periclymenum	18	35	6	52
* *Rubus fruticosus*	9	39	9	26
Rubus idaeus	9	28	0	29
Juncus effusus	9	0	0	0
Carex remota	0	3	0	0
Agrostris stolonifera	0	1	0	0
Frangula alnus	0	4	0	0
Poa nemoralis	0	3	0	0
Galeopsis tetrahit	0	8	3	0
Sorbus aucuparia	0	16	3	6
Quercus robur	0	16	6	6
Fagus sylvatica	0	12	12	16
Deschampsia flexuosa	0	0	0	3
Polypodium vulgare	0	0	0	6
Blechnum spicant	0	0	0	3

Fig. 9.8. A TWINSPAN dendrogram and cluster transition bar of the set of sequential relevés of the herbaceous transect area of the *Stellario-Carpinetum* 1978-1981-1985. *1*=*Oreopteris*, *2*=*Stellaria*, *3*=*Hedera* and *4*=*Lonicera* cluster. **B** Plotted scores of the relevés against the first two DECORANA axis. Relevés belonging to the same TWINSPAN cluster are delineated. Transition vectors that connect different year relevés of the same quadrat are drawn to indicate shifts of quadrats through the diagram. For reason of clarity of the figure, however, not all transition vectors could be drawn but only those that show characteristic shifts

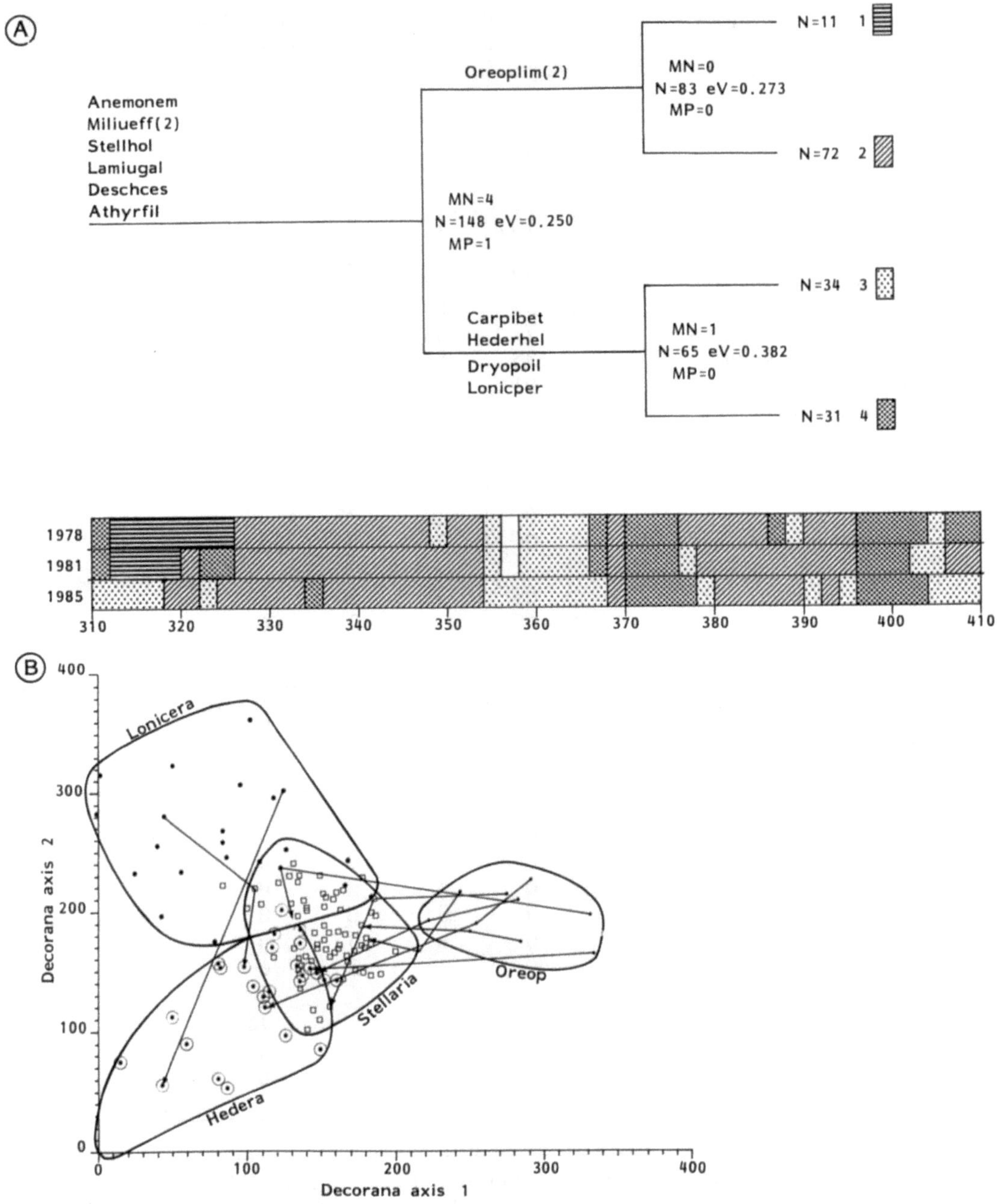
A
Anemonem
Miliueff(2)
Stellhol
Lamiugal
Deschces
Athyrfil
MN=4
N=148 eV=0.250
MP=1
Oreoplim(2)
MN=0
N=83 eV=0.273
MP=0
N=11 1
N=72 2
Carpibet
Hederhel
Dryopoil
Lonicper
MN=1
N=65 eV=0.382
MP=0
N=34 3
N=31 4
1978
1981
1985
310
320
330
340
350
360
370
380
390
400
410
B
Lonicera
Stellaria
Oreop
Hedera
Decorana axis 2
Decorana axis 1
0
100
200
300
400

on this axis are noted for the *Lonicera* cluster on acid substrate that is found on the woody debris, although the low values on this axis for the species-poor *Hedera* cluster do not allow these to be interpreted as less acid.

9.2.3 Conclusion

In conclusion, the clearest shift in species composition has been observed in dynamic gap eco-units that closed during the time period of the study. Differences in time can be best analyzed in sets of relevés where differences in abiotic site are slight. For example, the fourth cluster of the *Fago-Quercetum* transect interval, that has been interpreted as a spatial transition to the *Milio-Fagetum*, has no genetic relation to the other three clusters distinguished.

Chapter 10

Studies of Variables in Different Forest Areas

In this Chapter an arbitrary selection for comparison is made of variables that are available or can be computed from the forest model. The possibilities of deducing variables for different purposes are far from exhausted, and as yet have only been partly explored. Relations between parameters by which surface and volume can be expressed and which can be easily measured have been studied by many authors. Such relations play an important role in prediction models for forest structure development. Some of those relations between tree measures will be analyzed.

10.1 Crown Area Index and Crown Cover Index

Two measures of canopy density have been computed from the model with the program ARBOSUM (see Appendix).

First, the sum of projected crown areas of all life trees is computed and expressed as the total projected crown area above unit area of ground surface (Crown Area Index) according to Oldeman (1974a).

Second, the sum of projected crown areas multiplied with the inner crown cover per tree is computed and expressed as the real crown cover per area of ground surface (Crown Cover Index). These measures have been compared for a number of sites. In future, when records of the core areas are repeated, these measures can be used for comparing shifts in canopy density and in species composition (Table 10.1).

Table 10.1. Different core areas compared for the following values: Crown Area Index (*C.A.I.*) which is the total projected crown area per unit area of ground surface. The crown cover index (*C.C.I.*) is C.A.I. corrected per crown with the inner crown cover. The table also gives total crown volume (*Vcr*) and Basal area (*G*)

	C.A.I	C.C.I.	Vcr(m^3)	G(m^2/ha)
Bialowieza 256	1.645	1.022	105978	29.55
Fontainebleau I	1.434	1.055	126372	26.57
Fontainebleau II	1.583	1.118	130402	23.41
Ile de Rhinau	1.888	1.055	108444	24.75
Noordhout	1.437	0.942	83747	25.08
Otterskooi	1.349	0.802	56376	23.31

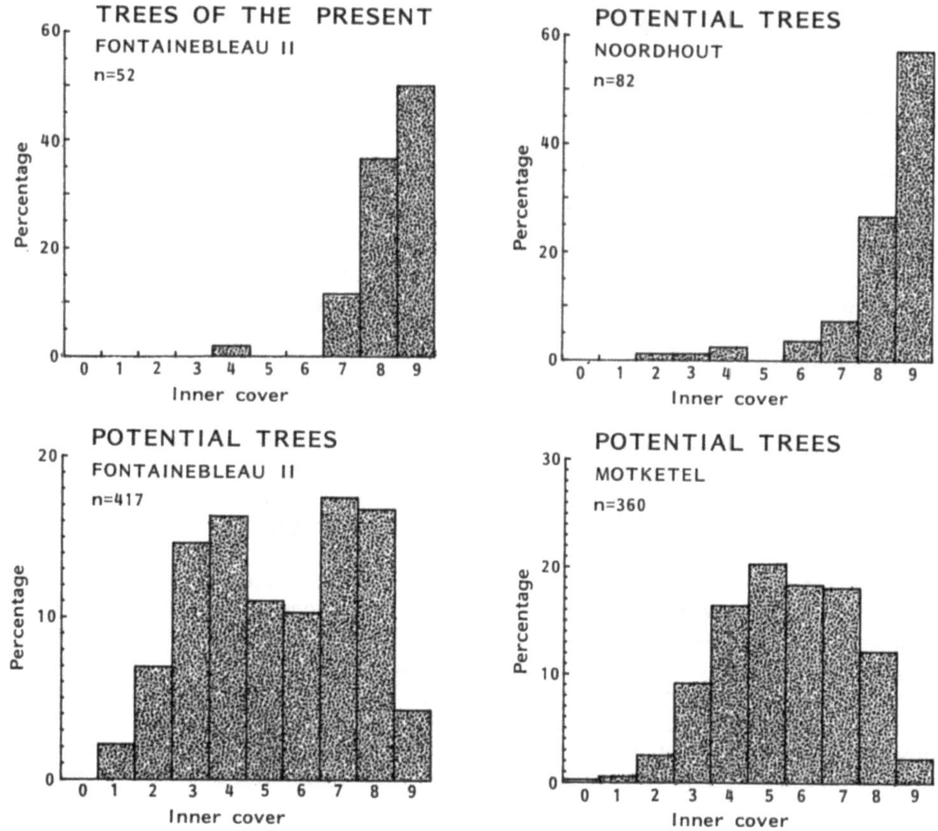

Fig. 10.1. Inner crown cover frequency diagrams of *Fagus sylvatica* in Noordhout, the Motketel (both The Netherlands) and in Fontainebleau II (France)

It is remarkable that the real C.C.I. is about the same for all the core areas. It does not matter what the values are of C.A.I., crown volume, basal area, the C.C.I. is about 1. Only the core area of the Otterskooi, which is an open fen woodland that passed a recent phase of decay (Chap. 7), has much lower figures for all variables.

10.2 Crown Cover per Tree Species

For numerous individual trees, the cover within the crown projection has been assessed. These data allow an analysis of the crown cover per tree species. Distributions of crown cover classes (see Appendix) per tree species will be compared. Differences for potential trees and trees of the present will be analyzed.

10.2.1 Crown Cover of Beech (*Fagus sylvatica*)

Suppressed shaded beech trees show lower inner crown cover than exposed trees. How the trees are distributed over the cover classes depends on the positions occupied by beeches in a certain forest build-up. This can be demonstrated by the beeches in the three core areas.

In the forest reserve Noordhout (The Netherlands), all beeches are potential trees that occur in low densities under a decaying, light-transmitting canopy of *Pinus sylvestris*. As can be seen in Figs. 5.2 and 5.3, the crown projections only overlap in their peripheral zone in the profile. Also the beech trees are of the same height and do not outshade each other. About 60% of all trees shows cover 9. The median value of distribution amounts to 9 (Fig. 10.1).

In the forest reserve of the Motketel (The Netherlands), beeches occur under a similar canopy of *P. sylvestris* as in Noordhout. The set of potential beeches, however, is much more heterogeneous in height and crown projection area. Many trees are outshaded by other beech trees. The range of inner crown cover classes in the frequency diagram (Fig. 10.1) is much wider. Only a few emerging trees show the highest cover and many trees in the shade show low inner tree covers. The median value of distribution amounts to only 6.

The forest reserve of Fontainebleau (France) is a pure beech forest. The set of the present as well as the potential set is occupied by beeches. The structuring trees belonging to the set of the present show a similar distribution in inner crown cover as the set of the potential trees in the core area of Noordhout (Fig. 10.1). The set of the present in Noordhout consists of light-transmitting *Pinus sylvestris*. The set of potential trees of Fontainebleau, however, is partly shaded by the set of the present of beeches and shows much lower inner tree covers. The distribution diagram shows two tops. One top, around the cover value 8, represents the potential trees that almost reached the upper canopy and occupy the relatively illuminated zones between the emergents. The second maximum, around the cover value 4, represents the suppressed trees in the deep shade below.

10.2.2 Crown Cover of Ash (*Fraxinus excelsior*) and Oak (*Quercus robur*)

The differences in cover distributions of the set of the present and the set of potential trees of *Fraxinus excelsior* and *Quercus robur* can be explained in a similar way as in *F. sylvatica*.

The most frequent cover class in the set of the present of *F. excelsior* in the core area of the Otterskooi is 7 (at the same time median value). The distribution diagram of the set of potential trees shows again two tops, but these are not as clear as in the beeches in Fontainebleau. The maximum around 7 represents the trees just underneath the upper canopy. The second maximum represents the numerous wilting treelets in the forest undergrowth (Fig. 10.2). The median value in the set of the present of *F. excelsior* in the core area of the Ile de Rhinau is 7. The median value for the set of potential trees is only 4 (Fig. 10.2).

The distribution diagrams of trees of the present and potential trees of *Q. robur* in the core areas of the Otterskooi and Noordhout are very similar to those of *F. excelsior* (Fig. 10.3). In the core area of the Norgerholt, only a set of the present of *Q. robur* exists. These trees have the same median value 7 as the trees of the present of *Q.*

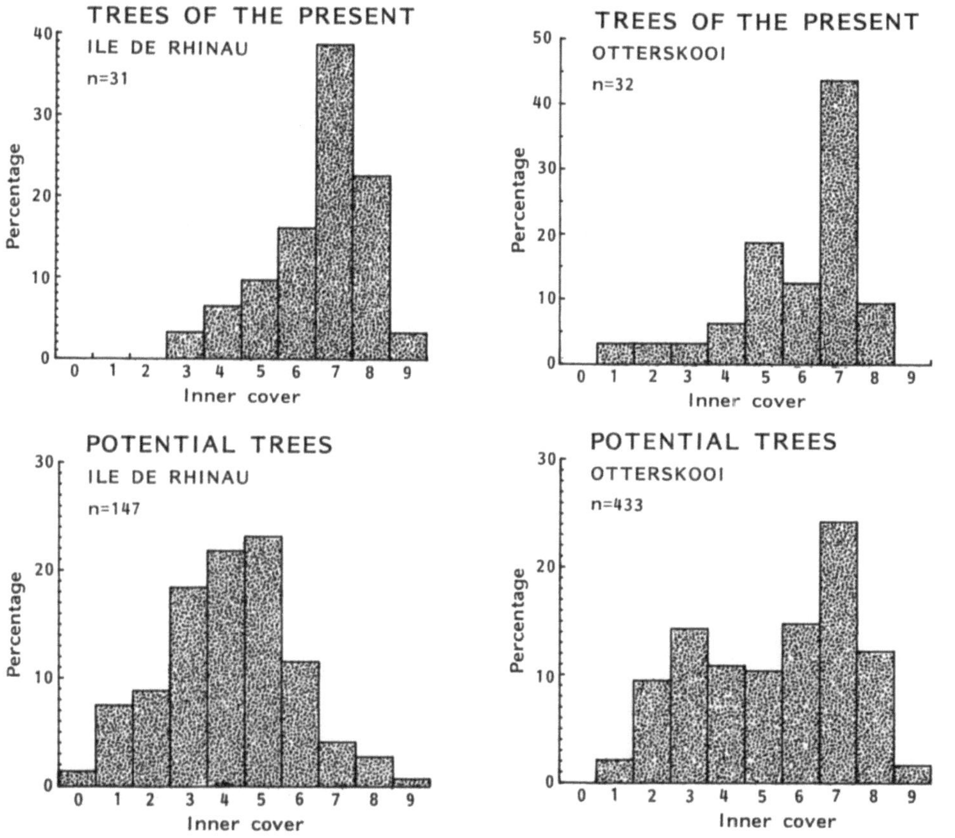

Fig. 10.2. Inner crown cover frequency diagrams of trees of the present and potential trees of *Fraxinus excelsior* in the Otterskooi (The Netherlands) and in the Ile de Rhinau (France)

robur in the Otterskooi and Noordhout.

Specific inner crown cover can clearly be distinguished in the set of the present. The trees of the present of *F. sylvatica* have the highest values of inner crown cover 9. *F. excelsior* and *Q. robur* both have a lower crown cover value of 7. In the more or less shaded set of potential trees, competition for light determines the inner crown cover. Shaded crowns develop an open mono-layered crown while crowns that have reached higher light levels develop multi-layered dense crowns.

10.3 Relation Between Diameter of the Crown and Diameter of the Bole

The relationship between crown diameter and diameter of the bole has been studied by, e.g. Ilvessalo (1950), Erteld (1979) and Dawkins (1963). Erteld (1979) for *Pinus sylvestris* reports a non-linear relationship that reaches near maximum for crown surface with increasing basal area.

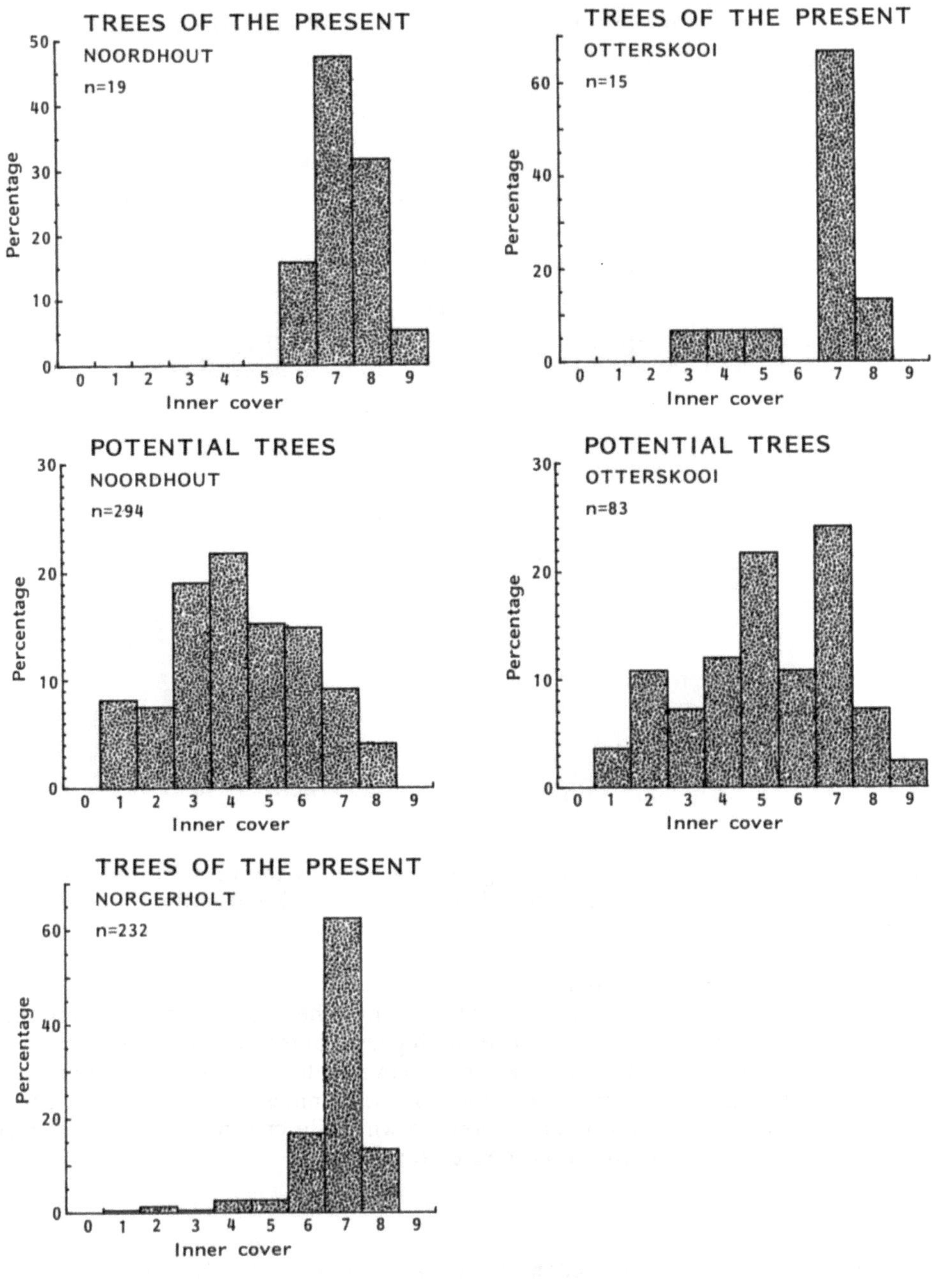

Fig. 10.3. Inner crown cover frequency diagrams of trees of the present and potential trees of *Quercus robur* in Noordhout, the Otterskooi and the Norgerholt (all The Netherlands)

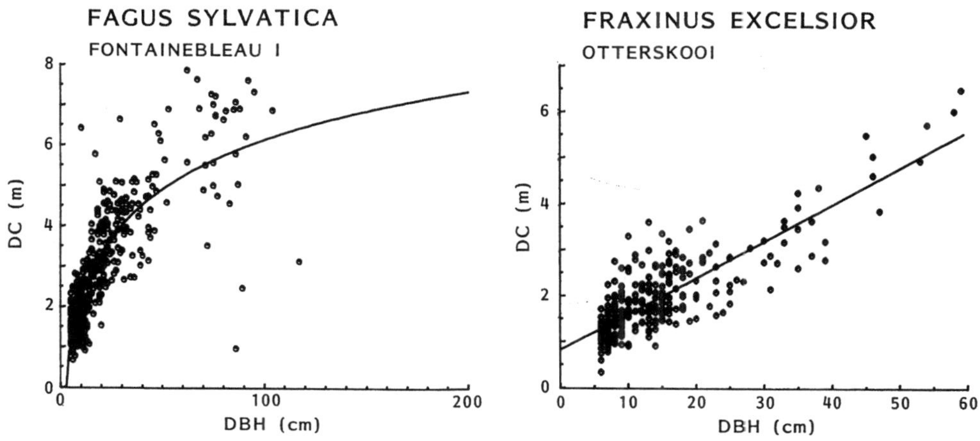

Fig. 10.4. Examples of regressions of diameter of the crown (*DC*) in metres against the *DBH* in cm. *Fagus sylvatica* proved to give best results with a logarithmic regression while in all other tree species linear regression gave best results

For most tree species linear regression provided the best fitting except for *Fagus sylvatica* in La Tillaie (Fontainebleau), where logarithmic regression proved better (Fig. 10.4). Although values for trees with a diameter at breast height < 5 cm were not measured, all C coefficients amount to about 1 m. This means that trees with a Dbh=0, which they have at the moment they surmount the breast height (1.3 m), already have a crown diameter of about 1 m. For some species the relationship between Dbh and crown diameter is very diffuse. Especially *Alnus glutinosa*, *Prunus padus* and *Pinus sylvestris* show values lower than 40% of the variance accounted for by the regression. This can partly be explained by the fact that many trees of *P. padus* and of *A. glutinosa* are bending. Because of bending of the tree crown the relationship may be distorted. By bending, the crown area has become larger, compared to an upright tree.
Trunks of the same girth of the different species carry crowns of different crown diameters. In Table IX.2 per core area the species have been sorted to increasing ratio of crown diameter against Dbh.

Pioneer species such as *Pinus sylvestris*, *Betula pubescens* and *Alnus glutinosa* have the lowest ratios of crown diameter against Dbh and at the same time a low percentage of variance accounted for. In all core areas they are in a starting phase of decline, which possibly may explain the diffuse relationship.

If the tree species with the best fitting of regressions (VAF > 50%) are sorted according to an increasing ratio of crown diameter against Dbh, we find the sequence *Picea abies*, *Quercus robur*, *Fraxinus excelsior*, *Carpinus betulus*, *Sorbus aucuparia* (Table 10.2). Any explanation for this fact has to be hypothetical as long as evidence is so scarse. One explanation might be found in the growth model (Hallé et al. 1978). Trees with a growth model with more frequent and spreading reiteration such as the model of Rauh in *Quercus robur*, and Roux in *Betula* spp. and Troll in *Fagus sylvatica* and *Carpinus betulus*, might have higher ratios of crown diameter against Dbh than the model of Massart in *P. abies*. Another explanation of the above-mentioned sequence may be found in the structural ensemble to which the trees belong. The trees that belong to the upper structural ensembles have the lowest ratios. Two tree species that definitely belong to lower structural ensembles such as *Carpinus betulus* in Bialowieza and *Sorbus aucuparia* in the Otterskooi show the highest A values. This might be seen as an adaptation to the relatively shaded regions in the forest, where they must spread their crowns for sufficient light interception (Terborgh 1985).

The ratio of crown diameter against Dbh in the same species shows big differences for trees in different core areas. For example, *Q. robur* in Bialowieza, Ile de Rhinau and the Otterskooi, respectively, shows increasing A values. So do the A values for *Fraxinus excelsior* differ for the Otterskooi and Ile de Rhinau respectively, but in a reverse sense. In Bialowieza *Q. robur* is surrounded and overtopped by *P. abies*, which can explain its low A value, while in the Otterkooi *Q. robur* belongs to the highest almost emergent trees and is not hindered by other trees from spreading its crown.
The different forest structure seems a logical hypothesis for the different A values although evidence is scarce.

Table 10.2. Linear regression coefficients *A* and *C* in Dc=AxDbh+C where Dbh is the diameter at breast height in cm and Dc is the diameter of the crown in meters. *N* is the number of trees. *VAF* is the variance accounted for in the regression. For *Fagus sylvatica* in La Tillaie (Fontainebleau) a logarithmic regression Dc=A x Ln(Dbh+1)+C provided a closer fit than a linear regression

Core area	Tree species	A	C	N	VAF
Bialowieza	*Picea abies*	0.0489	0.788	176	71.0
	Quercus robur	0.0589	0.875	13	87.8
	Carpinus betulus	0.0944	0.816	283	75.9
Ile de Rhinau	*Quercus robur*	0.0691	1.767	16	67.2
	Alnus glutinosa	0.0828	1.108	27	32.6
	Prunus padus	0.0870	1.158	63	16.3
	Fraxinus excelsior	0.0878	0.932	210	77.2
Otterskooi	*Alnus glutinosa*	0.0415	1.208	187	28.4
	Betula pendula	0.0650	0.703	41	47.6
	Fraxinus excelsior	0.0795	0.828	282	69.3
	Quercus robur	0.0801	1.110	82	44.2
	Sorbus aucuparia	0.1087	0.722	82	57.6
Motketel	*Pinus sylvestris*	0.0571	0.920	88	33.6
Noordhout	*Pinus sylvestris*	0.0609	0.586	111	43.4
La Tillaie I	*Fagus sylvatica*	1.7927	-2.11	435	72.8
La Tillaie II	*Fagus sylvatica*	1.9774	-2.352	468	77.7

10.4 Height/Diameter Ratios and Structural Ensembles

One of the most simple relationships is the ratio of height (H) and diameter at breast height (Dbh), studied, e.g. by Schulz (1960), Attiwill (1962), Cobble (1972), Nagano (1978) and Bücking et al. (1982, 1985). McMahon and Kronauer (1976) showed, following the principles of mechanical design, that trunk proportions are limited by elastic buckling criteria. Oldeman (1974a) and Hallé et al. (1978) established for tropical tree species the empirical rule of the relationship H=100d as a reference line. Trees, as long as they grow according to their architectural model, fit fairly well to this relationship. Trees with trunk regeneration of only one surviving new model just below the top of a broken trunk (bayonet-joint) overshoot temporarily and reach H>100d. Trees that entered a phase of profuse reiteration which eventually ends with herbaceous reiteration, such as trees in the canopy belonging to the set of the present, showed a height that was much less than 100 times their diameter.

The height-diameter reference line and cloud of points can be represented in a log-log graph (Oldeman 1974a, 1978). The shape of the cloud of plotted points, mostly ranging around the H = 100 x Dar (diameter above roots) line, gives information about the number of structural ensembles. A structural ensemble is the assemblage of all trees of the present that have their crowns in certain height levels, but defined by trees of the present only in one regeneration unit.

10.4.1 Height/Diameter Ratios and Structural Ensembles in Beech Dominated Forests of the *Fago-Quercetum* and *Melico-Fagetum*

Height/diameter ratios *Fagus sylvatica* were analyzed in two core areas, i.e. two semi-natural beech forests in Fontainebleau (France). The two core areas are situated on different site types. Fontainebleau I is a *Fago-Quercetum*, and Fontainebleau II is a *Melico-Fagetum* (see Chap. 7)(Koop and Hilgen 1987). Slender trees are most frequent in Fontanebleau I as well in the potential trees as in the trees of the present (Fig. 10.5 and Table 10.3).

Table 10.3. Height/diameter ratios in *Fagus sylvatica* in Fontainebleau I and II separately for potential trees and trees of the present. SE is the standard error of the mean. *STD* is the standard deviation

		Min	Max	Mean	SE	N	STD
Fontainebleau I	Potential	31	200	117	1.53	352	28
Fontainebleau II	Potential	30	200	101	1.48	445	31
Fontainebleau I	Present	28	121	63	2.27	84	20
Fontainebleau II	Present	25	183	57	3.62	57	27

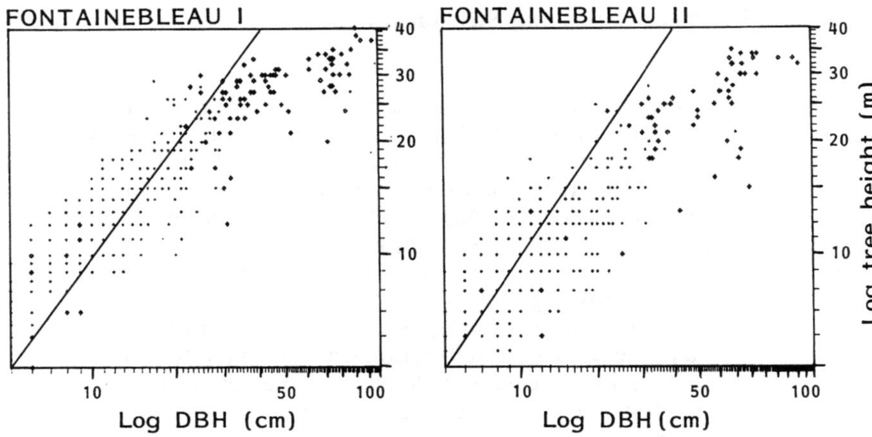

Fig. 10.5. Log Dbh against Log tree height of all trees in both core areas of Fontainebleau. The *line* of reference H=100d is drawn. Potential trees indicated with *crosses* and trees of the present with *small rectangular dots*

None of the other species in the two beech-dominated core areas reaches the state of tree of the present. In Figure 10.5 only beeches are represented in the set of the present, which in the diagrams can be identified by their low height-diameter ratio.

In the *Fago-Quercetum* plots *Betula pubescens*, *Sorbus aucuparia*, *Salix capraea*, *Populus tremula* and *Frangula alnus* occur as slender treelets, that try to reach the canopy but become overgrown by beech. In the *Melico-Fagetum*, *Fraxinus excelsior*, *Acer campestre* and *Carpinus betulus* share this fate. These species can only reach full dimensions in regeneration units without beech.

10.4.2 Height/Diameter Ratios and Structural Ensembles in Forests of the *Alno-Padion*

The biggest difference with the beech-dominated forests is that more than one species reaches the set of the present. The height-diameter ratios of *Fraxinus excelsior* have been studied on two different site types. In the Ile de Rhinau (Alsace, France) the site can be characterized as a periodically flooded *Fraxino-Ulmetum* along the river Rhine. The Otterskooi is a *Pruno-Fraxinetum* on floating peat soil. Height-diameter ratios of both sites again show differences. In the Ile de Rhinau height/diameter ratios are higher than in the Otterskooi (Fig. 10.6 and Table 10.4).

These two core areas of *F. excelsior* show differences in distributions of height-diameter ratios similar but far more pronounced than between those of *Fagus sylvatica* in Fontainebleau I and II. Obviously only on nutrient-rich sites is it possible for trees to grow up in shade towards a more or less closed canopy. Even the deep and narrow "ecological chimneys" (Oldeman 1974a) can be used by trees to climb up to the canopy. Young trees invest here in height growth first to reach more light and later, as

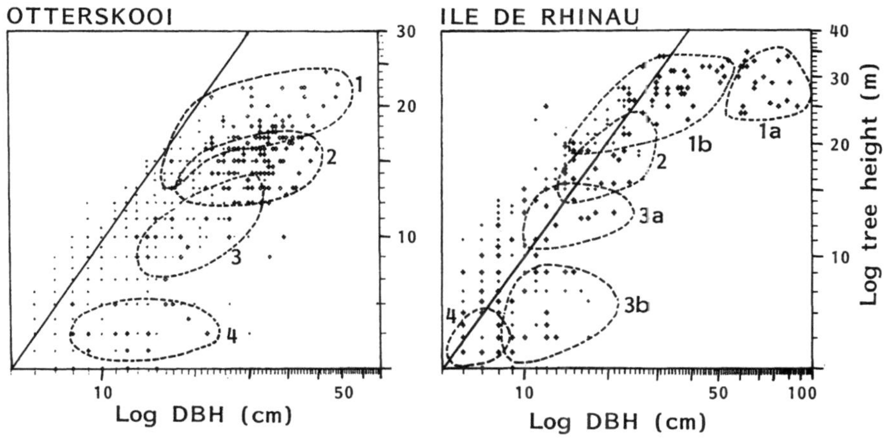

Fig. 10.6. Log Dbh against Log tree height of all trees in the core areas of the Otterskooi and Ile de Rhinau. Potential trees indicated with *crosses* and trees of the present with *small rectangular dots*. The *line* of reference H = 100d is drawn. Points of the trees of the present of one species are *delineated* and numbered. For the Otterkooi: *1 Quercus robur and Fraxinus excelsior*, *2 Alnus glutinosa* and *Betula pubescens*, *3 Sorbus aucuparia*, *4 Salix multinervis*, *Frangula alnus* and *Sambucus nigra*. For the Ile de Rhinau *1a Quercus robur*, *1b Fraxinus excelsior*, *2 Alnus glutinosa* and *A. pubescens*, *3a Prunus padus*, 3b layered *Prunus padus*, *4 Prunus fruticans*

Table 10.4. Height/diameter ratios in *Fraxinus excelsior* in the Otterskooi and in Ile de Rhinau separately for potential trees and trees of the present. SE is the standard error of the mean. *STD* is the standard deviation

		Min	Max	Mean	SE	N	STD
Otterskooi	Potential	10	70	48	1.29	145	29
Ile de Rhinau	Potential	25	250	96	1.42	431	15
Otterskooi	Present	11	70	46	2.99	29	34
Ile de Rhinau	Present	56	121	78	3.31	29	17

they have reached the canopy, they emphasize their diameter growth. The trees of the present that most recently became established in the canopy therefore show very high height-diameter ratios. On the nutrient-poorer sites, trees cannot grow up in shade because of decreased shade tolerance and need bigger gaps to reach the canopy. They show a "free-standing growth pattern" (Hallé et al. 1978, p.299, Fig. 84C,D). It is not

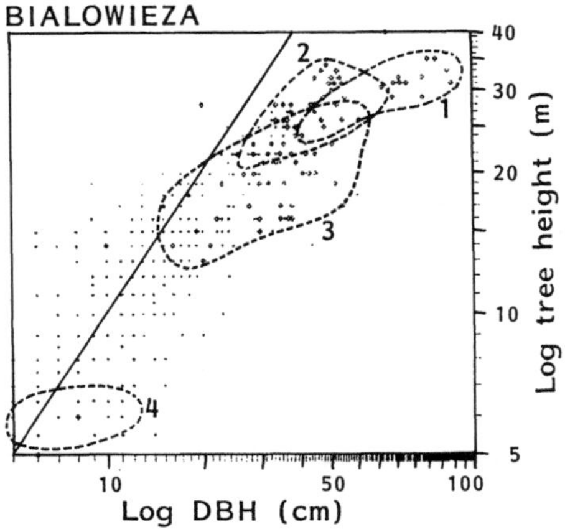

Fig. 10.7. Log Dbh against Log tree height of all trees in the core areas of compartment 256 in the National Park of Bialowieza. The *line* of reference H=100d is drawn. Potential trees indicated with *crosses* and trees of the present with *small rectangular dots*. Points of the trees of the present of one species are delineated and numbered. *1 Picea abies, 2 Quercus robur, 3 Carpinus betulus, 4 Corylus avellana*

neccessary to invest much in height growth first. In the open they have lower height-diameter ratios.

The log-log plots of Dbh against tree height of the Ile de Rhinau and the Otterskooi (Fig. 10.6) show a remarkable resemblance. The highest structural ensemble is formed by *F. excelsior* and *Q. robur*, although oaks in the Otterskooi are mostly potential trees, whereas in the Ile de Rhinau only trees of the present of oak exist and regeneration of the species fails. A second structural ensemble is formed by *A. glutinosa*, accompanied by *B. pubescens* in the Otterskooi and in the Ile de Rhinau by *A. glutinosa* x *incana*. A third structural ensemble consists of *S. aucuparia* in the Otterskooi and *P. padus* in the Ile de Rhinau. A fourth fragmentary structural ensemble can be distinguished in the Ile de Rhinau of *P. fruticans* and of *S. multinervis* and *Frangula alnus* in the Otterskooi. In the Ile de Rhinau a lower semi-structural ensemble is formed by uprooted but surviving *A. glutinosa* x *incana* and bended or layered *Prunus padus*.

Most tree species for example in the Otterskooi, show height/diameter ratios much lower than the H = 100d reference line proposed by Oldeman (1974a) for French Guiana. Best resemblance with this tropical and empirical relationship can be found in *F. excelsior*.

The regression coefficient of log overall height versus log diameter at breast height amounts to 90.4 (96 % variance accounted for). Potential trees are found above as well as below the regression line, while trees of the present only have ratios higher than to 90.4. Regression coefficients of *B. pubescens* amount to 81.1 (97 % variance accounted for), that of *Q. robur* to 79.2 (96 % variance accounted for), *S. aucuparia* to 77.7 (95 % variance accounted for) and that of *A. glutinosa* to 76.3 (97 % variance accounted for). In our data a line of reference as the H = 100d line for the tropics cannot be used for any site type and for any tree species.

S. aucuparia and young trees of *F. excelsior* that occur in the lower tree strata show the highest ratios of maximum height to diameter of individual trees of 143 and 133, respectively, followed by *Q. robur* with 105. These extremely slender trees primarily invest in height growth in the so-called "chimneys", i.e. narrow funnels or cylinders under a slight gap in the canopy through which the light energy level is raised (Hallé et al. 1978). Meantime these trees can remain erect because of their wind-sheltered position underneath the canopy. *A. glutinosa* and *B. pubescens* grew up in the open and at the moment when light-demanding regeneration underneath the canopy is almost lacking. The maximum ratios of these two species do not go beyond 98 and 100 respectively.

The log-log plots of diameter against tree height of two core areas in the sections 256 in the *Tilio-Carpinetum* in the National Park of Bialowieza (Poland) (see Chap. 7) show four structural ensembles. The upper ensembles overlap (Fig. 10.7). The highest ensemble is formed by emerging *P. abies* trees. The second by *Q. robur*, *T. cordata* and *A. platanoides*. The third ensemble is occupied by *C. betulus*. The lowest ensemble that can be distinguished is filled by *Corylus avelana*.

In conclusion, several structural ensembles exist in the associations *Tilio-Carpinetum*, *Fraxino-Ulmetum* and *Pruno-Fraxinetum*, the two latter belonging to the phyto-sociological alliance of the *Alno-padion*, and they spatially overlap.
On the contrary, only one structural ensemble is fully developed in the *Fagus sylvatica*-dominated forests of the *Fago-Quercetum* and the *Melico-Fagetum*. Lower structural ensembles, often referred to as shrub layers, here obviously do grow up in canopy gaps, but never reach maturity, since they are overgrown by *F. sylvatica*.

Chapter 11

Simulations

The importance of the height of the lowest living branch or fork, also called the morphological inversion point, is stressed by Ogawa (1965), Hallé et al. (1978) and Oldeman (1974b, 1989). It provides an easily measurable ratio between crown depth and total height. Empirically, Hallé et al. (1978) established that a ratio greater than one half indicates a free-standing, not shaded, tree. A ratio of about one half indicates trees that have recently entered a phase of vigorous growth. A value less than one half is evidence of diminished vigour, by suppression or by senescence. Hallé et al. consider the morphological inversion points collectively as the morphological inversion surface. This surface gives a topographical indication of the distribution of either vigorous, recently established, or senescent canopy trees. Lescure (1978) used the morphological inversion surface for characterization of secondary tropical rainforest development in French Guiana. Hallé et al. (1978) suggest that the position and nature of this inversion surface might have a bearing on microclimate regulation. Light and humidity gradients with height are modulated by the canopy architecture. These modulated gradients show microclimatic "inversions" from less or more than such light intensities as average extinction would account for at a certain level in the forest, and also from more or less than average humidity build-up. The collected points where these inversions take place can be thought of as forming an ecological inversion surface. Hallé et al. thus emphasize that efforts should be made to relate the morphological to the ecological inversion surface. When a constant relationship exists, determination of the one, which is a relatively simple matter, would give information about the other, without recourse to expensive micro-meteorological techniques. Hallé et al. (1978), according to personal communications and suggestions of Horn in 1974, discuss the hypothetical forest construction with so-called light "cells", that permit the development of tree crowns at lower heights below the canopy. Terborgh (1985) developed a similar model. He argues that the point-to-point variance of integrated light over a whole day in any given horizontal plane under the canopy is greatest just beneath the canopy. At this level any point is either directly under a crown and permanently shaded or directly under a gap. At greater depths below the canopy, a larger fraction of the space in any plane receives some direct illumunation. At a certain distance under the canopy the light fields of adjacent gaps begin to overlap. Still farther down, the light fields contributed by alternate gaps begin to overlap. At this level the point-to-point variance of integrated daily light field is nearly eliminated because any point in the horizontal plane receives light from two or more gaps. If trees use the higher light values between the canopy trees, this means that they become irrevocably committed to one gap. Terborgh gives two reasons why a midstorey species such as *Cornus florida* in the south-eastern United States does not become committed to one canopy gap. It expands its crown at an early stage, deviant from canopy trees. As the species has reached the height of several metres, the crowns are already broader than a typical canopy gap. Continued growth upward would thus entail a counterproductive reduction in crown area. A second reason for not becoming committed to one canopy gap is the dynamic character of these gaps if compared to the life-span of the species of Individual gaps are an unreliable light source in the long term.

11.1 Light Plans and Profiles

11.1.1 Horizontal Light Patterns

Direct radiation patterns and diffuse radiation patterns at ground level are different (Figs. 11.1 and 11.2). Contour lines that enclose areas with the highest values of direct radiation generally show stretched oval forms with a lengthways west-east orientation. Areas with low values of direct radiation show more rounded forms. Radiation values computed for points directly at short distance north of stems are much lower than the surrounding radiation values. This is caused by the local effect over a very resricted area of shading by the tree stem. By interpolation this area is exaggerated. The stretched oval forms of direct radiation can be explained by the east-west tracks of the sun. The light-beam coming from the sun that follows an east-west track and passes through a canopy gap causes sunflecks on the forest floor to move in an opposite west-east track. In the pattern of summarized light, values over more than a day of these west-east tracks of sunflecks cause the oval stretched forms of high values of direct radiation. From the collation of direct radiation pattern with the crown projections map, it is clear that the areas with highest values of direct radiation have shifted to the north in comparison to the canopy gaps because of the oblique south-inclined sun radiation. They are situated at the northern edges of bigger canopy gaps, while in case of small gaps the effect reaches under the closed canopies just north of the gap.

The lower the sun tracks are, due to latitude or season, the bigger this shift will be. With diminishing inclination of the sun, the effective canopy gap will change its form and dimensions and at lower inclination it will disappear. Only sunshine within a restricted period around noon causes the areas of high direct radiation.

Lower or bending trees have direct local effect on direct radiation, which can disrupt the pattern described above. In the Otterskooi (Fig. 11.1) the areas with lowest values of direct radiation are confined to the *Sorbus* unit and to an area under an uprooted oak tree that still lives. The *Sorbus* unit (Chap. 7.5) is characterized by a low structural ensemble of relatively dense crowns of *Sorbus aucuparia*.

The pattern in diffuse radiation has less steep gradients. The darkest places are circular and are situated directly under low tree crowns. The effect of canopy gaps reaches in all directions underneath the closed canopy (Fig. 11.2).

11.1.2 Vertical Light Patterns

The pattern of direct radiation has also been simulated in a vertical plane. It has a whimsical pattern caused by tree crowns just behind and in front of the plane. Dark areas can be found where this plane dissects a tree crown. Any differences in computed radiation values inside tree crowns are the result of differences in irradiance from outside the tree crown, because the model assumes that tree crowns are just transparent bodies without any inner crown differentiation in leaf densities. Tree crowns high above the forest floor are isolated dark islands, while the low crowns spread their shade down to the ground.

In the upper regions of the canopy the extinction is greatest; the countour lines are closest to each other. The figures confirm the theory of greatest variance in

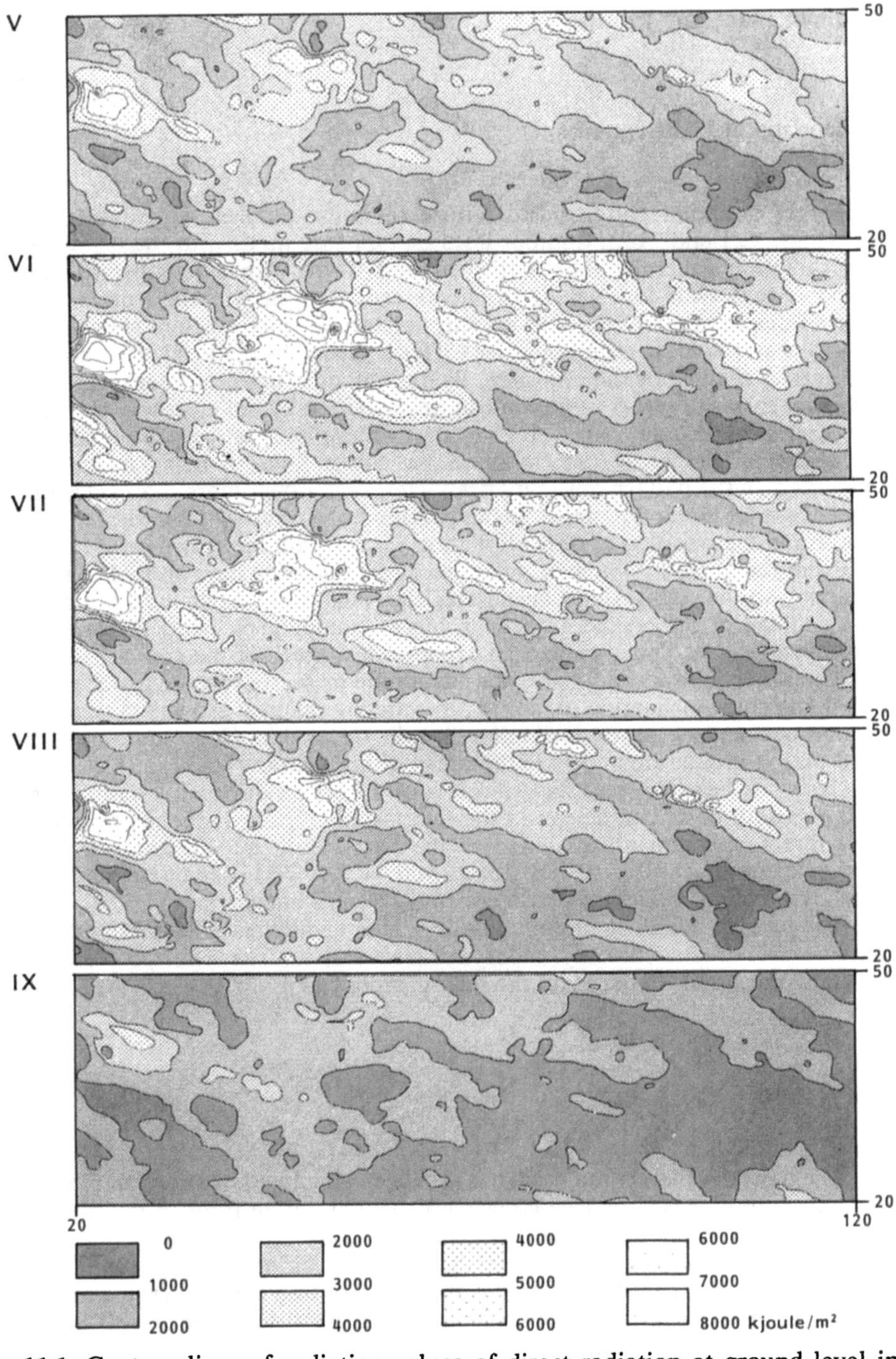

Fig. 11.1. Contour lines of radiation values of direct radiation at ground level in the central 100 x 30 m block of the core area of the Otterskooi as computed by the program FOREYE over the months May-June-Juli-August-September (*V* to *IX*).50-100 years

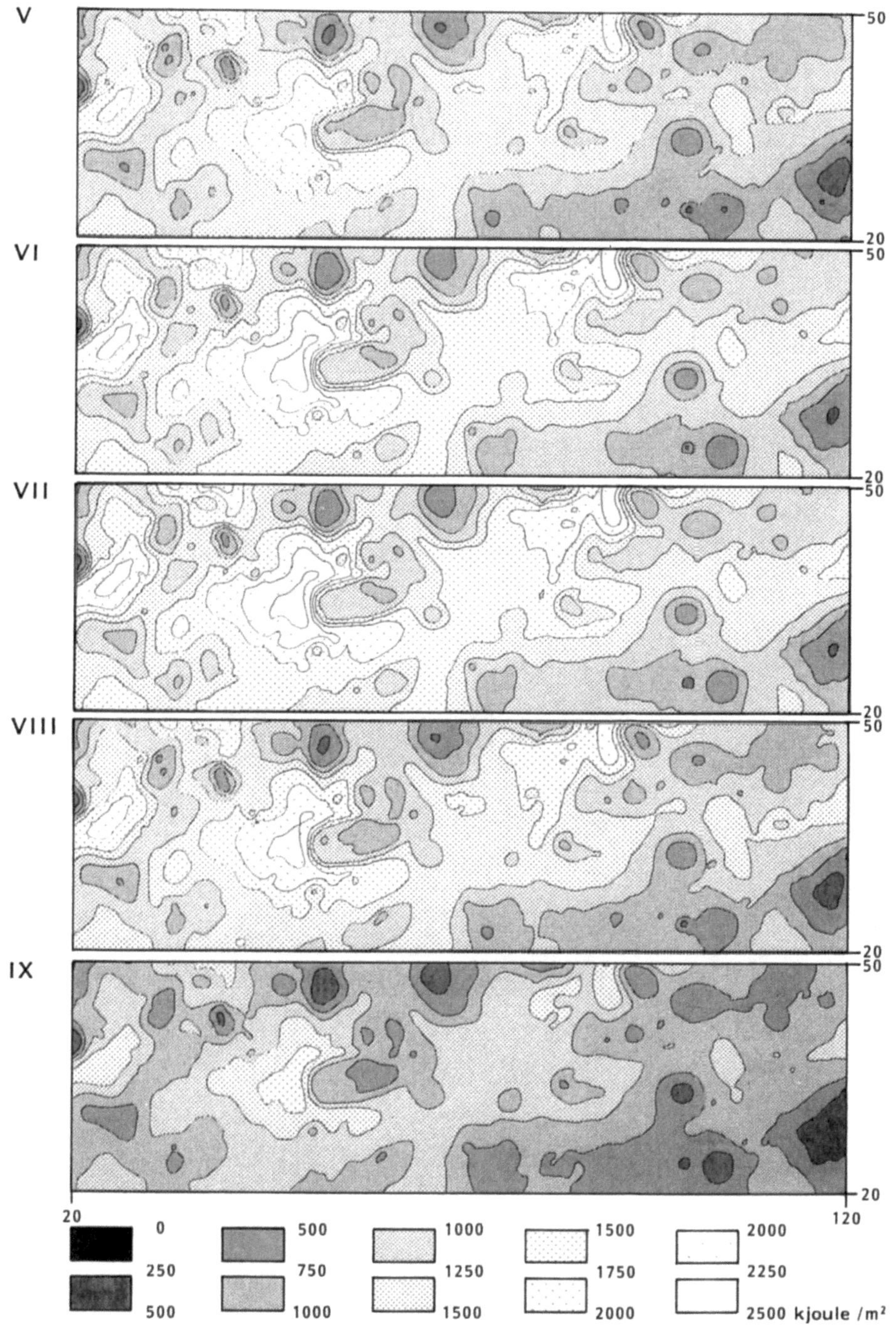

Fig. 11.2. Contour lines of radiation values of diffuse radiation at ground level in the central 100 x 30 m block of the core area of the Otterskooi as computed by the program FOREYE over the months May-June-Juli-August-September (*V* to *IX*).

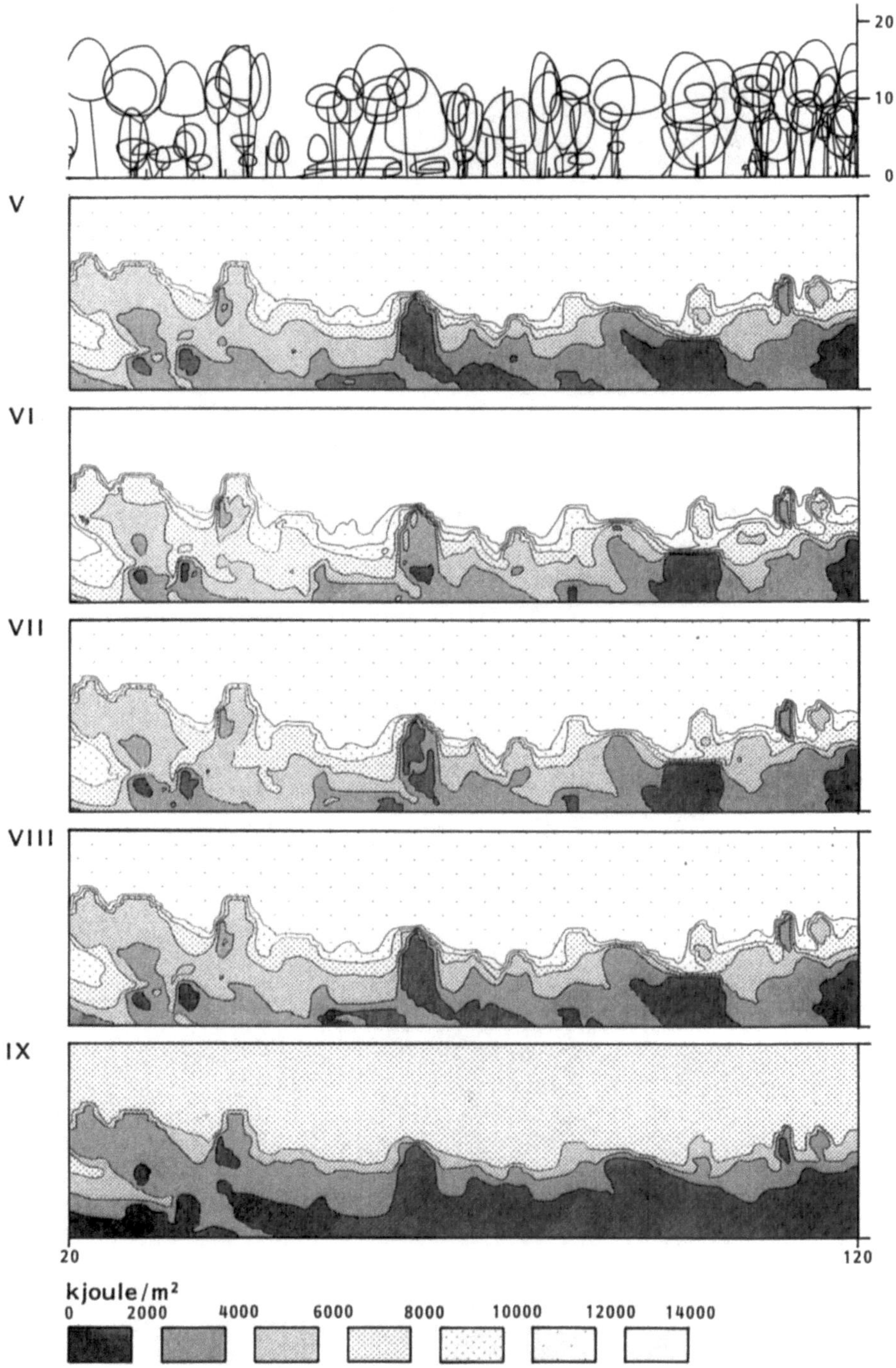

Fig. 11.3. Contour lines of radiation values of direct radiation in a vertical plane at the middle of the woody species transect area of the Otterskooi as computed by the program FOREYE over the months May-June-Juli-August-September (*V* to *IX*).

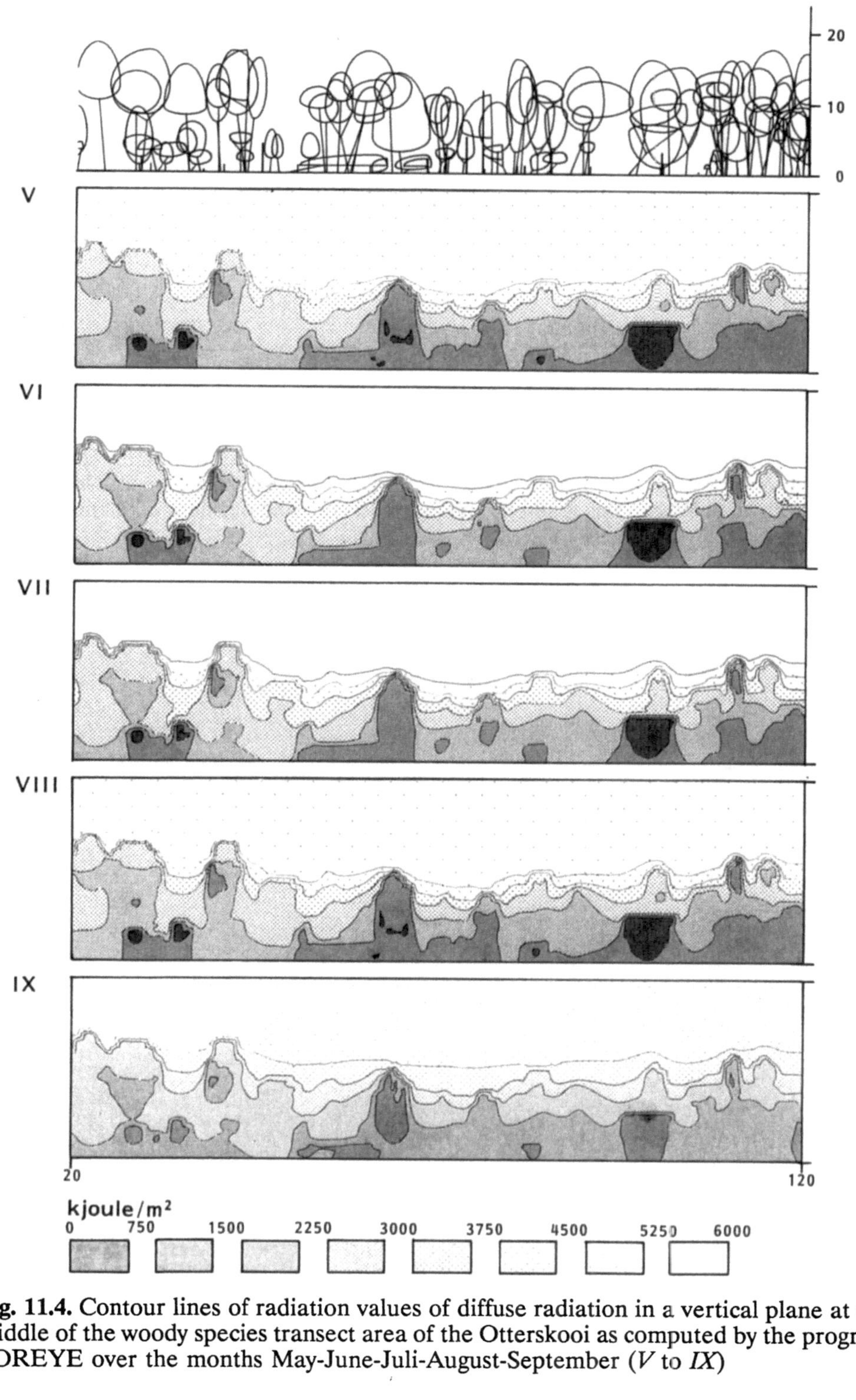

Fig. 11.4. Contour lines of radiation values of diffuse radiation in a vertical plane at the middle of the woody species transect area of the Otterskooi as computed by the program FOREYE over the months May-June-Juli-August-September (*V* to *IX*)

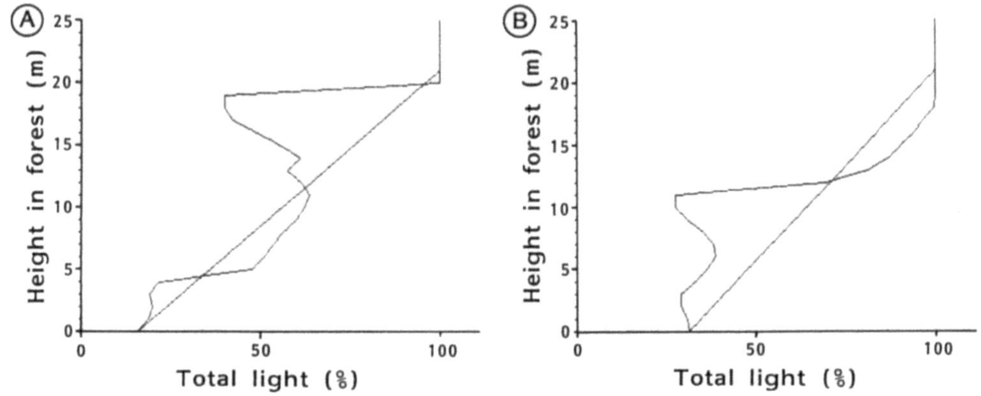

Fig. 11.5. Gradients of direct radiation as computed by the program FOREYE over the month of June at two perpendicular lines (**A** and **B**) in the transect area of Noordhout (The Netherlands). Where this gradient line intersects the line of linear extinction points of inversion can be found

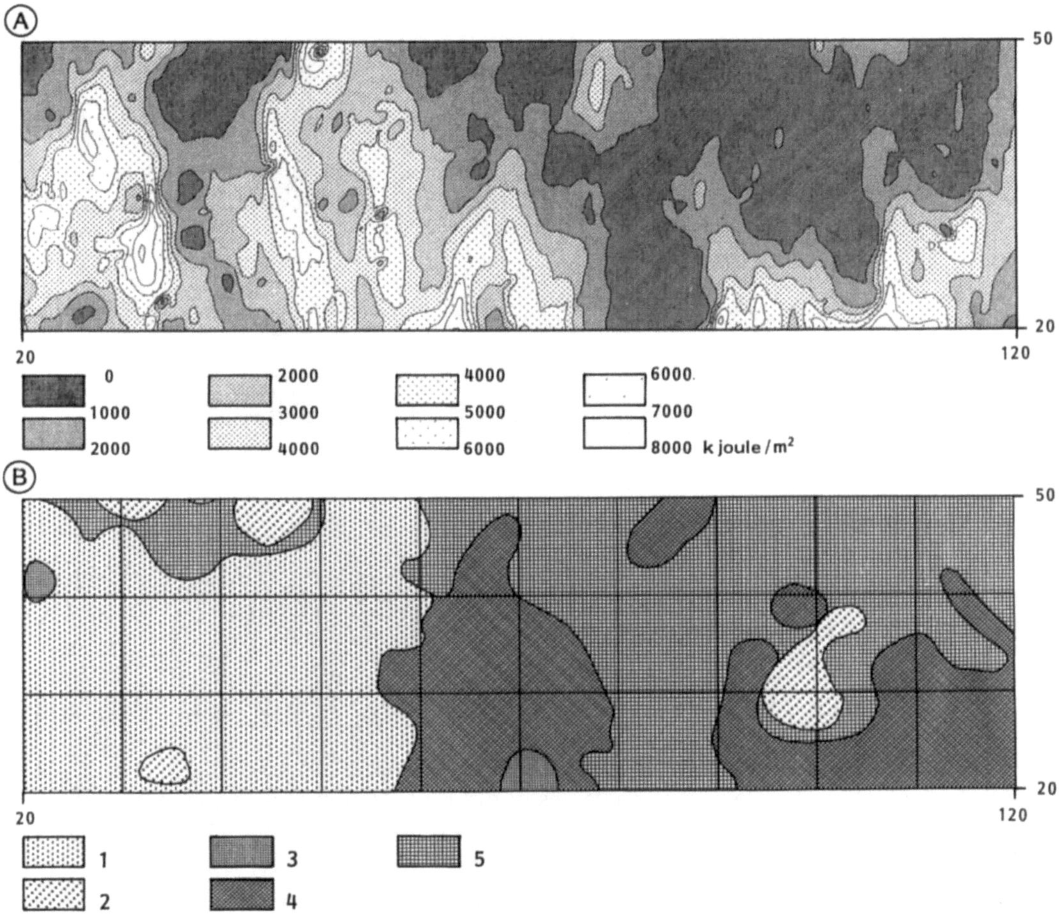

integrated light just below the canopy of Terborgh (1985). In Figure 11.5 radiation gradients show the ecological inversion points (sensu Hallé et al. 1978) mentioned above. The smaller and the more interactive the eco-units, the more the inversions are hidden under complex shapes of curves caused by oblique radiation. The light patterns can be collated with herb species occurrence in the herbaceous transect and with growth of trees in the future. Collations of the light pattern of direct radiation in the month of June with the vegetation map of the core area of Noordhout (Fig. 11.6) shows the relation between areas with radiation values < 1000 KJ/m^2 and the occurrence vegetation types 2 and 5. Both vegetation types are differentiated from the other types by the abcence of *Deschampsia flexuosa.*

◄

Fig. 11.6. A Contour lines of radiation values of direct radiation at ground level in the central 100 x 30 m block of the core area of Noordhout (The Netherlands) as computed by the program FOREYE over the month June. **B** Vegetation pattern of the same area. The legend units of the map are named after dominant and codominant species. *1* = *Vaccinium myrtilus-Vaccinium vitis idaea-Deschampsia flexuosa* type, *2* nude forest floor, *3* = *Vaccinium myrtilus-Vaccinium vitis idaea* type, *4* = *Vaccinium myrtilus-Deschampsia flexuosa* type, *5* = *Vaccinium myrtillus* type [illegible] (Tan[illegible] 1985). [illegible] earthquake inversion points [illegible] et al. 1989) [illegible] and the more interactive the area [illegible] the lowest [illegible] [illegible] [illegible] as caused by oblique [illegible]. The [illegible] [illegible] [illegible] to [illegible] occurrence in the hereafter [illegible] and a [illegible] [illegible] of the fluid [illegible] [illegible] [illegible] of the [illegible] area of [illegible] (Fig. 11) [illegible] [illegible] values [illegible] 2000 K [illegible] [illegible] [illegible] the other type [illegible] [illegible].

[illegible] values of [illegible] (Netherlands) [illegible] the month time, [illegible] [illegible] [illegible] [illegible] [illegible]

Part D The Potential of SILVI-STAR

Chapter 12

The Monitoring System as a Tool: Some Examples

In this Chapter we will look back and compare the developed system with its objectives (Chap. 1).

An operational method for description of forest dynamics is now available. The 58 forest objects (see Appendix) of the RIN forest ecological information system, of which some have been used in this book for demonstration, have been under study now for 9 years. Full reassessments of the oldest core areas will be made in 1990. The reassessments will provide data on mosaic shifts in the core areas such as could be worked out earlier for the core areas of Fontainebleau (Koop and Hilgen 1987) due to a link up with the work of Bouchon et al. (1973).

The method has been adopted and used for 2 years by workers of the state forest department in describing the initial state of ten forests incorporated in the Dutch forest reserve project (Bosch 1986). Collected data are stored in the ORACLE database system as described in this book. In this connection the method is part of a planned interdisciplinary research program in Dutch forest reserves that embodies collaboration of institutions, studying among other things soil and litter dynamics and natural regeneration and development processes.

With the development of cartoon film techniques and the possibility of plotting and image processing techniques related to geographical information systems, data on forest dynamics can be visually transmitted even to persons who are quite unfamiliar with the dynamic character of forest development.

The gap between the descriptive methods and the practice of forest management on the one hand and the forest modellers on the other can be bridged if modellers made use of the same data sets. Preferably, growing data sets of monitoring projects should be used, so that models and their results can be tested against observed forest dynamics. Monitoring and modelling should go hand in hand. Often, simplification of the data gathered in the information system presented is necessary before it can be used as input for a specific model. The same data set, however, could serve for other models which do include more information on local single tree forest structure and can produce images and figures which are closer to and more recognizable for the practical management people. Examples of the potential use of the system for different disciplines will be given below. Also the use for consultancy and extraction of data for other purposes than originally planned will be presented.

12.1 Forestry

Although time series are too short yet to draw gross conclusions concerning the questions put forward in the problem analyses in Chapter I, some preliminary results can demonstrate the use for forestry practice. Conclusions about competition between trees can be drawn for different forest types. The monitoring of the spontaneous process of self-thinning indicates whether or not management is necessary to reach a certain goal. For example, in the *Melico-Fagetum* and the *Fago-Quercetum* in Fontainebleau, no other species than *Fagus sylvatica* plays a lasting role in a spontaneous forest development. This means that other species will not succeed unless they are continuously released from obtrusive beeches. It is swimming against the stream. At the same time it was found that within 1 ha core area a small-scale uneven-aged regeneration mosaic of beeches can exist. These findings indicate that a management system of beech with small-scale cuts and harvest of tree groups could run with minimal correction of the natural process and thus would be most cheap.

The core area in the *Tilio-Carpinetum* of Bialowieza, however, shows that at this site type a mixture of tree species can exist. A certain gap size is necessary for pioneer species such as *Quercus robur* or *Populus tremula* to regenerate.
For each forest type the critical gap sizes for different tree species to succeed can be found. Such knowledge can be used to stimulate natural regeneration of specific species. The duration of different phases in a developmental cycle on each site can be used for designing thinning systems. Forest management regimes on similar sites can mimic the scale and sequence of forest dynamical phases and with minimal corrections a mixture of tree species can be maintained (Brünig 1981).

A first step towards a predictive tree growth model has been made by the light simulation of FOREYE. Until now the light values have been used only in a correlative way with tree ring increment and reaction of herbaceous flora and tree regeneration.

For interpretation of remote sensing images, whether they be aerial photographs, scanning or radar images, the three-dimensional forest model is a useful tool to extrapolate the groundtruth data collected in the core area to the surrounding forest. The model could be used for canopy reflectance modelling according to, e.g. Li and Strahler (1985, 1986), Franklin and Strahler (1988), Goel and Grier (1988).

12.2 Fauna Habitat Studies

Forest structure plays an important role in the habitat selection of many forest animals. Measures like the number of tree species per unit area, cover and height of herbs, shrubs, and trees have proved to determine habitat selection. Foliage height diversity (MacArthur and MacArthur 1961), basal area and tree diameter distribution were important as well (Mühlenberg 1980; James and Wamer 1982). Kneitz (1980) and Van der Plaat (1982) made preliminary studies of what species groups would be suitable for monitoring inside the core areas of forest reserves. The impact of forest structure upon species has been studied most frequently in birds (e.g. Karr and Roth 1971; Willson 1974; Blondel and Cuvillier 1977; Opdam and Van Bladeren 1981; Opdam and Schotman 1986) and small mammals (e.g. Olszewski 1968; Deuser and Kaufman 1978; Deuser and Shugart 1978, 1979; Jonge and Dienske 1979; Vickery 1981; Dankers 1983 and Piepers 1986). August (1983) and Van Vuure (1986) studied the effect of forest structure on other species groups.

Methods of forest structure description for fauna habitat description as mentioned by James and Shugart (1970), van Berkel (1979) and Mühlenberg (1980), can all be deduced from the SILVI-STAR digital forest model. In fact, the three-dimensional model of crown volumes distributed in space enables calculation of new variables for characterization of habitat complexity and heterogeneity.

Microclimate is an important factor in habitat selection of tropical birds (Karr and Freemark 1983). For insects such as ants microclimate is the determining factor for nest location (Mabelis 1977). Stocki (1981) in a preliminary study described insect life-spaces as spatial volumes determined by forest architecture in strip transects. Akker and Groeneveld (1984) tried to recognize architectural insect habitats in the Amazonian rainforest. In this context the radiation program FOREYE could offer new possibilities of monitoring fauna habitat selection. Other fauna habitat requirements could be deduced from the SILVI-STAR model, such as spatial conditions for nesting, shelter and food availability.

12.3 Pollution Impact

The information system will notify changes in vegetation composition and vitality of tree species due to pollution (De Wit 1982). Soil acidification from atmospheric ammonium sulphate in forest canopy throughfall in the areas with intensive animal husbandry may affect the vegetation (Van Breemen et al. 1982). Because it is impossible to build climate chambers for the ecosystem level without excluding natural stress factors such as storms, unpolluted blancs with "normal" levels of forest dynamics are lacking. Therefore it is difficult to separate the pollution impact from a "normal" rate of tree mortality. Probably the lethal level of the "normal" stress factors, e.g. extreme dry seasons, cold winters, late frosts in spring, is reached sooner. Trees are more prone to diseases and phytophage insects by weakened immune response (De Kam and Van Dam 1987; Schneider and Bresser 1988; Schütz and Van Tol 1989).

In The Netherlands an increase of a number of nitrophytic species such as *Corydalis claviculata* and *Moehringia trinervia* could be observed even during our relatively short period of data collection. This increase must probably be ascribed to the increased nitrogen deposition from intensive animal husbandry. Increase of nitrophytic species or species that can take up N at high N/P ratios at the expense of other species has been observed in grasslands and heathland (Heil and Diemont 1983; Van Dam et al. 1986; Bobbink and Willems 1987).

12.4 Nature Management

For species conservation, the knowledge of niches in spontaneous development cycles is indispensable. In which forest types can species find their niche in spontaneous forest development ? How long does a whole forest developmental cycle take, and in which periods within that cycle can species find their niche ?

The examples of Fontainebleau with 600 years of spontaneous forest dynamics has revealed the very long-term effect of clear-cuts upon generation changes of trees and development of the silvatic mosaic. Planned natural areas that were previously managed forests and are to be left unexploited can be expected to show similar lag effects. Transformation management of the artificial forest structure can anticipate the mature patterns described for different site types (Koop 1986). From a comparison between the multispecies *Fraxino-Ulmetum* and the *Pruno-Fraxinetum* forests of the Ile de Rhinau and the Otterskooi, it is clear that different growth rates of different species and a faster growth on nutrient-rich sites can accelerate fragmentation of an originally homogeneous stand.

Since forest grazing is regarded as a tool for nature management in forests (Van Wieren and Borghesius 1988; Van Wieren et al. 1989), the observations in the old pasture woodlands in the New Forest gain in interest. Valuable results of these transects can be applied in consultancy concerning the prediction of effects of forest grazing. Because of the shifting mosaic character of forests, besides the forested phases, lawns, shrubs, and fringe vegetation play an important role. For example, a species-rich lawn vegetation which succeeded the collapse of the central old Denny Wood and the role of shrub encroachment and cage effects of fallen tree crowns on different site types in the New Forest are interesting facts that can be translated into potential development on similar site types. It could be proven that it is a myth that forest grazing breaks up homogeneity in no matter which even-aged forests.

12.5 Hydrology

The three-dimensional SILVI-STAR forest model can serve as a tool in modelling the spatial aspect of interception, throughfall and stemflow in heterogeneous forests. Aerodynamic roughness of the canopy influences through its impact on transpiration rates the interception capacity of the canopy (e.g. Thom et al. 1975; Rauner 1976; Jarvis et al. 1976; Veen 1984) The number of tree crowns to be passed before water reaches the ground influences throughfall (Mitscherlich 1971; Olszewski 1976; Bouten and de Water 1987; Bouten et al. 1989). Falinski and Olszewski (1977) found a clear relationship between mean snow cover thickness in the course of one winter and canopy coverage in the Bialowieza National Park. Stemflow depends largely on tree species (Geiger 1961).

12.6 Soil Studies

Radiation is an input in models of litter turnover (e.g. Bouten 1984). In studies on the spatial heterogeneity of litter and humus profiles in the field situations, the large variation of irradiance values in space and time in a forest and the time-consuming nature of radiation measurements are a handicap. The FOREYE radiation model would serve to quantify spatial heterogeneity of radiation input. This application of the SILVI-STAR model will be elaborated in future in the multidiscipinary forest reserve project.

12.7 Vegetation Science

Long-term observations on forest dynamics in unmanaged forest will help to understand the nature of the vegetation in general, the Potential Natural Vegetation in particular. By the monitoring of forest dynamics it will become clear which part of spatial heterogeneity can be explained by dynamic aspects of one and the same PNV. Definitions and description of forest types in terms of PNV can be supplemented with knowledge about forest dynamics, such as duration of developmental phases, species composition of those phases, e.g. as described for the Neuenburger Urwald (Chap. 9) and the availability in time of specific microhabitats such as mound-and-pit micorelief, or dead wood that serves as a niche to special plant species.

The FOREYE radiation model can help specifying niches of forest plant species in regard to shade tolerance. Questions about what are the critical phases for establishment of species and limiting levels for maintenance in regard to shade tolerance can be solved in a number of cases.

Chapter 13

Scope for Improvements and Further Development

The detail and precision of description of the monitoring system has been worked out in an iterative process of feedback between description and the level of explanation required for management purposes.

The system hierarchy for description was attuned to the scale of management and the practical questions that had to be answered. This hierarchy allowed causal explanation of pattern and processes at one hierarchical scale by pattern and processes at adjacent system levels. In this chapter improvements will be discussed that could result in better prediction. Adaptations of the tree model and the application programs, as steps towards the predictive model, are proposed. Further application programs to be developed are mentioned.

13.1 Restrictions of the System

Although in an early phase of development a preliminary study was carried out for the incorporation of the fauna in the monitoring system (Van der Plaat 1982), a complementary monitoring project on fauna was never started because of shortage of manpower. The department of Landscape Ecology of the Research Institute of Nature management carried out some breeding bird inventories in forest site areas under study with the monitoring system. These data have not yet been fully elaborated.

The abiotic component is missing in the monitoring system. Although incorporated in an earlier study (Koop 1981), no manpower was available for soil characterization in the present monitoring system until now.

In a multidiciplinary study project in strict forest reserves on the initiative of the State Forest Service (Bosch 1986) the University of Amsterdam and the Research Institute for Forestry "De Dorschkamp" will collaborate to enable incorporation of abiotic and fauna components in the information system.

Other research groups, attracted by a detailed multidisciplinary environmental description of a potential study site, participate in monitoring other forest components. For example, in all forest reserves fungi will be assessed during 3 successive years in the woody transect area (Veerkamp 1989). Similarly in some reserves mosses and lichens will be monitored. On the maps of the core area each tree can easily be retraced individually and the simulation of the light climate with the program FOREYE offers new possibilities for description of their habitats.

13.2 Prospects for Further Development

13.2.1 Data Collection

The reassessment of the set of forest objects under study will be continued. New forest reserves will be incorporated in the monitoring program as part of the Dutch forest reserve program. As time series of the permanent plots of the herbaceous transect area grow (now maximal 3 reassesments) and the full reassessments after 10 years of the core areas become available, these will offer new possibilities for description of forest dynamics.
Other data sets from abroad will be processed in the same way, such as those of repeated transect analyses in the montane rainforests of Java (Indonesia) by Smiet (1989).

13.2.2 Predictive Models

First of all the program FOREYE has to be validated with real radiation measurement in the field. Until now these labour-intensive measurements were beyond the scope of the monitoring project, which primarily focussed on continuation of reassessment of the forest objects under study. In the next few years first attempts at validation will be carried out.

As a continuation of the line set by the "cartoon film"-forest growth reconstruction by the program GROWRING, in future we will try to develop a predictive forest growth model that simulates single tree growth using the data of the monitoring system for calibration. The prediction should focus on relatively short time runs, over time periods that are relevant for management purposes. Its main use is in training courses for demonstration of the effects of decisions for silvicultural design. The program FOREYE provides simulation of the competition for light. However, for this use the program needs adaptation (see below).

Possibly routines for other competitive factors could be included such as the competition model of Faber (1983). This model was originally designed for mono-species and mono-layered stands and presumes a more or less closed canopy. It takes into account the individual dimensions of trees and their spatial arrangement, to compute competition factors and simulate individual tree growth. Competition is approximated by the growth space that is assigned to a tree. The total of growth space is divided in a grid. The pixels are assigned to individual trees according to their weight factor. The weight factor is the higher the shorter the distance of the tree to the pixel and the thicker the tree. In Faber's model the concept of horizontal growth space is intended to cover aboveground competition as well as root competition. However, it is evident that in uneven-aged vertically complex forest architecture its relevance for root competition is higher than for aboveground competition. Therefore for simulation of root competition, the model of Faber could be added complementarily to the FOREYE procedure, which takes into account the aboveground competition for light.

The risk of death or windthrow could be simulated in the three-dimensional forest model (see below). However, at this moment it seems premature to expect realistic predictions over long time periods, because of the stochastic character of tree death and the damage caused by tree falls. These long-term predictions probably find a better solution in the JABOWA like gap models at the cost, however, of spatial relationships among individual trees and among regeneration units.

The program CANOPIX to compute the upper canopy surface still suffers from the fact that light intensities reflected from a rough canopy surface show large variation. This largely depends upon the provenance of light reflection, either from shaded, embedded deeper parts of the canopy surface or from higher exposed or sun-enlightened parts. Shading pattern should be simulated, so that for the training pixel samples shaded pixels of the same species can be separated from the unshaded ones. This extra information could be taken into account when interpreting training sets of reflection values of pixels in two or more colour bands.

A new application of the SILVI-STAR forest model is probably found in interpretation of radar scanning images. Because the radar waves of different wavelengths and polarization penetrate into the vegetation at diferent rates (e.g. Hoekman 1985, 1987), the three-dimensional forest model can help with the interpretation, especially of complex forest signatures on non-light images, in respect to species composition and structure.

13.3 Problems to Solve

For incorporation in a predictive tree growth model the present single-tree three-dimensional tree model needs some adaptation.

A first problem to be solved is that an iterative use of the radiation simulation procedure (the present FOREYE), such as would be needed in a predictive model, is rather time-consuming. The time needed for one point computation of FOREYE depends on the maximum tree height and the tree density. The higher the point for which radiation is computed, the more trees can be skipped for computation and the shorter is the computing time. The maximum computing time for one point now is 5 CPU min at a VAX 750 computer in the case of a 30-m-high forest with a tree number of 1000 per ha. The computer time could be reduced by increasing the size of the steps in which the tree crowns are scanned. In the present version of FOREYE the step size is one degree. Another option is simplification of the present complex crown section. For this specific purpose, future trials will be run to simplify the tree shapes of compound quarter ellipses and vertical crown curves to the surfaces of revolution. Sensitivity tests of the program FOREYE will have to be carried out with different step sizes for scanning and simplified tree shapes. If simplified tree shapes, such as surfaces of revolution, which need less computing time, result in comparable light values these will be more suited for iterative radiation simulation for prediction of forest growth.

A second problem to be solved if we want to keep open the opportunity of asymmetric crown development is the difficulty of altering the form of the horizontal crown section during tree growth. In the program GROWRING we let the compound of quarter ellipses shrink without altering its shape.

In the first-mentioned case, a change in shape of the horizontal section of the tree crown should depend on the radiative conditions at different parts of the tree crown. In this case, the relative differences in radiative conditions on the tree crown or, for example, in the four peripheral points, must determine which path the four peripheral points should follow to build a new compound of quarter ellipses. The surface area to which the new horizontal section can grow is, however, limited and determined by competitive factors. Probably a simplified tree model, as a surface of revolution, is the best solution to start with.

The distribution of risk of windthrow in a forest is an interesting phenomenon in regard of the frequencies and dimensions of disturbance.
The three-dimensional forest model could be used for modelling stability of individual trees against windthrow (Ford 1976; Kuiper and Van Schoten 1985; Kuiper in prepraration). Not trying to simulate the very complex reality of turbulent wind flows, the risk factors could be determined for windthrow or stem break. Roughness of the canopy can be computed from the forest model and wind speed profiles could be simulated. Exposition of an individual tree embedded in the surrounding forest canopy architecture can be computed from the forest model. Together with the height/diameter ratio, the indications of the dimensions of the root systems per tree species obtained from uprootings and the dimensions of the tree, its crown and inner crown cover, a relative risk factor might be calculated.

Summary

A comprehensive monitoring system has been developed inclusive of a computer package for analysis of long-term forest dynamics observations. The system name **SILVI-STAR** is a composition of **SILVI**genesis and **S**ingle-tree **T**hree- dimensional **AR**chitecture.

The Need to Develop SILVI-STAR

On a worldwide scale there is the need for expertise on how to incorporate natural processes in forest management systems. Whether we deal with man-made forests or with semi-natural or natural forests, to ensure forest preservation and lasting production capacity, natural processes must not be pushed into the background by economic or technical progress, but incorporated in forest management systems. Knowledge of the natural processes serves not only silviculture but also management in nature preservation. At the present natural forests are often exploited without any care for the durability of resources, and for short-term economic profit natural processes are ignored.

Often man-made forest are managed much like an agricultural crop, at a high cost of regeneration. Especially in man-made forests that have matured, spontaneous developments should be exploited, so that for forest maintenance less intensive human intervention is necessary. To understand the dynamic character of forest development and the role of natural processes, long-term observations are necessary. However, only few data sets on time series of permanent quadrat survey, repeated forest stand measurements or forest profile analyses exist.

Most existing predictive forest models relate to the growth of no other species than trees, whether in stands or in patches that are assumed to be homogeneous. Studies of relations between trees and non-arborescent forest compartments are mostly lacking. Simulations of complex forest architecture of natural forest is still beyond the reach of any forest model.

The way of starting with a simple model that can later be elaborated in more detail is rejected. Instead a descriptive model of the complex reality has been developed and time series of forest development at different sites were built up. On the basis of the data collected, applicative computer programs were developed to solve specific problems on the way towards a predictive model.

Development of SILVI-STAR

In developing the monitoring system three phases have been described: first, the working out of appropriate techniques for data collection in the field; second, the organization of data storage and retrieval; and third, data processing and interpretation. The system was worked out from already existing techniques for the analysis of structure and species composition of natural and semi-natural forests, viz. (1) remote sensing, (2) forest profile analysis, (3) methods of forest inventory used in natural forest reserves in Central Europe, (4) permanent quadrat survey in combination with vegetation mapping and (5) ground photography.

Processes in the forest take place at different scale levels in space and time. The hierarchy theory (Simon 1962), which is concerned with systems that have a certain type of hierarchically organized complexity, has been the basic concept in the system design. Because the first visible impact of most silvicultural treatment is structural, forest structure has been chosen as the prime criterium in delimitation of four hierarchal levels. These levels: the tree compartment, the regeneration unit, the non-woody compartment and the mosaic of regeneration units correspond to the scale levels that are most relevant for management purposes.

The Nested Plot Data Collection

Taking into account the advantage of standardized metric plots and the natural limits and minimum areas of the hierarchic units, a method of nested plot data collection on forest architecture and plant species composition has been worked out.

I The forest type area The forest type is an area of 10 to 40 ha around the core area (II). A forest history map is made at scale 1:2500. Earthworks are mapped and the age of the woodlands is mapped as reconstructed from old maps and written documents. The spatial patterns of vegetation composition and of vegetation structure according to developmental phases are recorded. The forest type area is mapped from large-scale aerial photographs (1:4000 to 1:8000) every 10 years. Aerial photographs in true colour, colour infrared, or both, are digitized by means of a video camera or scanner and image files are kept on tapes. Additionally, visual photo interpretations, such as vegetation maps and forest structure maps, are kept on shrink-proof tracing sheets.

II Core area A rectangular 70 x 140 m core area, ranging one tree height around the 10 x 100 m woody transect area (III) is measured at the same time intervals. Trees bases, crown projections, uprootings and dead trees on the forest floor are charted at scale 1:200 for trees with a Dbh more than 5 centimeters. Of these trees, diameter is measured at breast height (Dbh). Heights are measured for the crown top (HT), the greatest width of the crown or the periphery height (HP), the crown base (HC), the first living fork (HF), and the relative height of the tree base to a reference height (HB).

All trees are classified according to their species, architectural code according to Oldeman (1974a), vitality, and cause and intensity of damage according to Leibundgut (1959, 1982) and Mayer (1976). The cover of the crown within its projection is estimated for the individual trees with the decimal scale of Londo (1984).

The position and stage of decay of dead wood is recorded. For part of the core area, tree ring samples are collected for all trees with a Dbh exceeding 10 cm. At fixed and marked camera points, every 5 or 10 years depending on the rate of change of the particular forest vegetation, black and white photographs (negatives 60 x 60 mm) and colour-slides (24 x 36 mm) are taken.

The vegetation pattern is mapped at scale 1:200. The core area is used as a groundtruth reference for the forest type area. Information on forest structure and dynamics from the core area are used for explanation of changes in floristic composition in the central herbaceous transect area consisting of 50 adjacent 2 x 2 m permanent quadrats (IV), that are assessed every 3 to 5 years.

III The woody transect area The tree layer of a permanent transect measuring 10 x 100 m is characterized, in sufficient detail to describe tree shape, with profile drawings and crown projections at scale 1:200 at 10 year intervals.

VI The herbaceous transect area The herbaceous transect area, split up in adjacent 2 x 2 m quadrats, enables analysis of species composition within the map units of the 1:200 vegetation map of the core area.

Data are collected on field-forms or as hand drawn maps that are later digitized. The use of a hand-held computer in combination with a tree caliper and a hypsometer enabled direct data input into files in the field. To guarantee continuity in data storage and query, a commercially available database (ORACLE) and a geographical information system (ERDAS, ARC/INFO) have been used in the construction of the information system.

The Tree Model

A three-dimensional single-tree architecture model has been worked out to describe asymmetric tree shapes with a minimum of measured data points. The volume occupied by one tree at a certain moment in a certain forest stand is described by the co-ordinates of eight points, representing: tree base, place of first fork, top and base of the tree crown and the four extremities (peripheral points) of the crown circumference in positive and negative x- and y-direction. All z-co-ordinates are derived from height measurements while x- and y-co-ordinates are digitized from the hand-drawn plan. X- and y-co- ordinates of the top and base of the crown and of the fork, that are not drawn in projection, are computed separately. By making use of the ellipsoid nature of most deciduous trees, a best-fitting wrapping is constructed by connecting the four peripheral points of the crown circumference with a set of quarters of ellipses.

The crown curves of vertical transverse sections through the top and the base of the crown and the peripheral points are represented by an equation which is cognate to the formula of the quadratic equation of the ellipse but has a varying exponent E. With help of the hand-drawn transect profile, a best- fitting exponent is approximated, per species, for the upper crown part by measuring the crown radius at two different heights. The E-value for most broadleaved tree species is about 2.

To prevent distortion of bending trees, with their position of the tree base situated outside the crown projection, these crowns are digitized and processed in a rotated co-ordinate system. The new x-axis is stretched between the stem base and the most remote point of the crown, in relation to the stem base.

The position of fallen dead trees in the model is indicated as a straight or bifurcated line. Bifurcations of a higher order are omitted. The position of uprooting is indicated as a compound of quarter ellipses. The four extremities in x- and y-direction of the mound in horizontal projection are digitized in a rotated co-ordinate system, with the x-axis parallel to the midline of the trunk, similar to a bent tree.

Time Series

Time series of forest development at different sites are built up on the basis of the digital descriptive model of the complex reality of forest structure and species composition.

The collected data provide a detailed, yet comprehensive, database enabling descriptive and explanatory modelling at different scales for forestry, vegetation science and fauna habitat studies.

From the ORACLE database any type of data or combination of data types can be selected. By using the program ORACLEX, a complete standard format file for one object can be retrieved from the database that fits all computer programs developed for data processing. The programs exist in VAX-FORTRAN for data processing and plotting on the computer (VAX and micro-VAX) and in BASIC for the Përsonal Computer (Olivetti). A visual interpretation of data is enabled by graphical system output such as profiles and ground plans of tree crown projections, providing substitutes for traditional profile drawings and maps. Crown area, crown cover, crown volume and basal area can be computed with the program ARBOSUM. These values are also computed separately for species and 10 x 10 m blocks of the core area.

Application Programs

In an application program (CANOPIX), for integration with remote sensing studies, an aerial view of the forest canopy is simulated on the basis of measured plot data. This view provides a groundtruth reference for the interpretation of remote sensing images. To explain the growth of individual trees and the distribution patterns of herbs and tree regeneration on the forest floor, another application (FOREYE) was developed, simulating the penetration of direct and of diffuse light. The three-dimensional forest model allows computation of the hemispherical cover for fixed points, inside the forest and above soil level, as an indication of the interception of diffuse light and changing oblique radiation. Cumulative cover values can be computed for the separate angles of the hemisphere from all crowns of trees within a certain radius. Light reflected by leaves, stems or branches is not taken into account.

The resulting values of such computation are processed with the subprogram FORFLUX in a way similar to a fish-eye lens- or hemispherical photograph (Anderson 1964; Ducrey 1975; Bonhomme 1976). The underlying radiation model is based on standard functions of direct and diffuse light under clear sky and standard overcast sky in J/m/s against sun height (de Wit et al. 1978). The distribution of diffuse light over brightness zones of the hemisphere is determined with the formulas of Ross (1981), distinguishing clear and standard overcast sky. Dependent on the hourly values, the chance of sunshine at any moment is determined. There is an option for random or fractionary determination of hourly sunshine duration. The amounts of direct and diffuse light can be computed over any specified period of the year. However, because the data on inner crown cover have reference to summer assesments in the period May-August, only computations in this period are reliable. The fish-eye computations for radiation assessment are made for the 50 adjacent relevés of the herbaceous transect area or in a grid in the ground plan or profile to derive horizontal and vertical light gradients as determined by the forest structure.

For the reconstruction of forest growth with tree ring data, a technique of animation was elaborated, facilitating a visual interpretation of the forest development.

The system is applied to demonstrate forest development in some European forest reserves using forest architecture descriptions and vegetation relevs, tree ring data and historical sources.

The Forest Reserve La Tillaie in Fontainebleau (France)

In the forest reserve of Fontainebleau (France) 600 years of forest development could be reconstructed for two core areas. Probably more or less even-aged stands became established after a cut documented in 1372. They decayed after they had reached their physiological maximum age as described in a document in 1664. The present forest is an almost pure beech (*Fagus sylvatica*) stand. The oldest of the two present tree generations that can be recognized in the tree ring data dates from 1651-1693 and 1609-1723 in the two core areas, respectively. Then in both core areas until 1850 there is a regeneration gap. A second documented phase of decline of the almost pure beech forests started after 1850. From that moment on, trees of the second generation established themselves almost continuously until 1950. Until the present day in both plots part of the old generation survived and regeneration continued, which means that this second phase of decline is lasting for a longer time than the first one in the seventeenth century. The generation change has spread in time and in space. Nevertheless, the effects of the cuttings in the fourteenth century can still be seen in the present forest structure.

When looking at the pattern of canopy height of both core areas in Fontainebleau, it is striking that the lower trees are spatially separated from the big trees. They are arranged in circles around the big trees of a biostatic patch, showing a network woven between the highest trees.

The National Park of Bialowieza (Poland)

In the National Park of Bialowieza (Poland) forest structure could be traced back for more than a century. During the eighteenth and nineteenth centuries, big *Picea abies* trees dominated a large part of the transect area. Some parts were dominated by *Carpinus betulus* and a big *Quercus robur*. Under the canopy of those broadleaved trees *P. abies* trees became established at the end of the eighteenth century. Some of those *P. abies* trees now have overgrown the *C. betulus* canopy. Around 1917, many big *P. abies* trees were uprooted. This event triggered the establishment of three groups of *Populus tremula* trees, of which one group together with an *Acer platanus* tree, can be found in the middle of the transect. During the natural thinning in *P. abies* during the late 1940's, in the first part of the transect a new generation of *P. abies* became established. Until 1984 a gradual decline of *P. abies* and *P. tremula* took place. The canopy gap formed by the death of an old *C. betulus* in 1943 was filled by a regeneration unit of mainly *Tilia cordata*. The young *T. cordata* trees at the margin of the opening show lateral crown growth in the direction of the light. The storms of 1984 and 1985 uprooted the remaining emergent *P. abies* trees.

The New Forest (England)

From historical data and comparison of the actual patterns of forest vegetation and recent developments in ten transects in the New Forest general conclusions about mosaic development in pasture woodlands have drawn. Development of a mosaic of forest phases depends on the coarseness of the mosaic with which grazing starts. If a fine-grained mosaic is the starting-point, long periods of "overgrazing" retaining generation changes will reduce age differentiation and result in fusion of formerly uneven-aged forest patches into open areas.

In general, grazing will retard spontaneous fragmentation of uneven-aged forest stands unless grazing pressure differs within a relatively short distance or fluctuates in time. Differentiation in grazing pressure at a short distance can be induced by complexity of the environment. Complexity of the environment can be produced by obstructing lying dead trees and crown cages and scattered canopy gaps where establishment is triggered and growth of trees is faster than under a homogeneously closed canopy, with few slowly growing young trees. The chance of survival of at least some young trees in such complex environments is greater than in homogeneous environments.

Ile de Rhinau (France)

A pioneer *Alnus-Salix* forest on a sediment alluvial soil, historically documented in 1880, soon after establishment became invaded by *Quercus robur* and *Ulmus* spp. A second regeneration wave of *Q. robur* and *Ulmus* spp. and the first establishment of *Populus* spp. trees around 1880 indicate the starting collapse of the pioneer forest. In the 1930's and 1940's *Fraxinus excelsior* invaded the forest. Only a few older *F. excelsior* trees were found in the core area. At that time, the *Salix-Alnus* forest had probably collapsed, while in the meantime the river dynamics declined due to the construction of the Grand Canal d'Alsace. Although the forest structure suggests an uneven-aged structure, three distinct tree generations can be recognized from the tree ring analysis. Different tree species with their own growth rate and maximum life-span resulted in a complex forest structure. In comparison with the core areas of Fontainebleau, it is striking that due to more than one tree species and various but higher tree growth rates, generation changes take less time and the homogeneous regeneration unit of *Salix* and *Alnus* fragmented much quicker.

The Otterskooi (The Netherlands)

The forest dynamics of the Otterskooi have been reconstructed in another way, i.e. without using such tree ring data as were available for the forests dealt with in the previous paragraphs. The reconstruction has been based upon interpretation of historical data and present structural features such as spatial distribution patterns of tree species, patterns in architectural compartments and tree diameter distribution. Because of a serial succession process relation with ground vegetation has been stressed.The patterns in the vegetation map and in the forest structure map could be explained from forest history.

Tree species distribution has been worked out from digitized aerial photographs using the groundtruth of the core area. The first generation forest shows the highest amount of *Betula pubescens*. In the older forests their share decreases. The portion of *Alnus glutinosa* slightly decreases towards the oldest forest parts while the amount of *Quercus robur* and *Fraxinus excelsior* increases, gradually forming coherent patches. In the youngest forest parts only a few isolated small *F. excelsior* and *Q. robur* crowns were located.

In the core area, forest history and potential development has been analyzed using the present data of forest structure in relation to the vegetation pattern in the core area such as diameter distribution diagrams. The distribution patterns of all tree species are separately plotted for two different architectural compartments, i.e. the set of potential trees and the set of trees of the present. A third graph is made of reconstructed crown projections of the trees of the past.

The role of *A. glutinosa* and *B. pubescens* declines. Trees of the present of *A. glutinosa* still indicate the dominant role of the tree in the present situation. Potential trees of *A. glutinosa*, however, play a role of some importance only on the younger, thin peat soils of the ditches. *Q. robur* first established itself on the oldest peat soils, that first became dry. Later establishment of the species in vegetation types with younger soils took place, as these places had become dryer.

F. excelsior is the most frequent potential tree in all vegetation types. In its whole range of occurrence, *F. excelsior* has the highest proportion in potential trees. *F. excelsior* that was first present in the so-called *Impatiens* type later invaded the dryer types. On the youngest peat soils and the ditches and away from the open water the role of *F. excelsior* as a potential tree declines.

The forest development deduced from the present forest architecture and species composition has been collated with succession theories in literature. Together the patches of the Potential Natural Vegetation of *Carici elongatae- Alnetum*, *Pruno-Fraxinetum* and the *Lysimachio-Quercetum* form a fine-grained site mosaic which is highly determined by the pattern of ditches and "ribs" and the distance to flowing water. The hierarchy of the sylvatic mosaic as a subsystem of the site mosaic has been interchanged here. The forest architecture has different potentials at a short distance, which results in a heterogeneous and diverse forest structure and species composition. Because succession has not ended yet in a steady state, it seems likely that the patches of *Carici elongatae-Alnetum* will develop into a *Pruno-Fraxinetum* or a *Lysimachio-Quercetum* in the future.

Monitoring

Tens of years or even centuries of observation are necessary to detect forest dynamics as expressed by architecture and tree species composition (see Chap. 7). However, fast initial changes in structure and species compósition can be exampled in young, spontaneous willow stands in Zuidelijk Flevoland, a polder reclaimed in 1968 (The Netherlands). In 3 years the mortality of *Salix triandra* and *Salix alba* x *fragilis* was much higher than of *S. alba*.

To detect changes in the herb layer composition, long sequences of relevés of the 2 x 2 m herbaceous transect area quadrats would be needed. Because the Neuenburger Urwald was subject to an earlier study in 1978 and was included in the monitoring system later, data on the Neuenburger Urwald have been used for an interim analyses.

In conclusion, the clearest shift in species composition has been observed in dynamic gap eco-units that closed during the time period of the study. Differences in time can be best analyzed in sets of relevs where differences in abiotic site are slight.

Studies of Variables in Different Forest Types

Some variables have been compared in different forest core areas under study with the monitoring system. Relations between parameters by which surface and volume can be expressed and which can be easily measured have been studied by many authors. Such relations play an important role in prediction models for forest structure development. Some of these relations between tree measures will be analyzed.

It is remarkable that the C.C.I. (crown cover index) is about the same for all the core areas. It does not matter what the value are of C.A.I. (crown area index), crown volume or basal area, the C.C.I. is about 1. Only the core area of the Otterskooi, which is an open fen woodland that passed a recent phase of decay, has much lower figures for all variables.

Specific inner crown cover can clearly be distinguished in the set of the present. The trees of the present of Fagus sylvatica have the highest values of inner crown cover 9. Fraxinus excelsior and Quercus robur both have a lower crown cover value of 7. In the more or less shaded set of potential trees competition for light determines the inner crown cover. Shaded crowns develop an open mono-layered crown while crowns that have reached higher light levels develop multi-layered dense crowns.

Height/diameter ratios indicate structural ensembles. Several structural ensembles exist in the associations *Tilio-Carpinetum*, *Fraxino-Ulmetum* and *Pruno-Fraxinetum* but they spatially overlap. Whereas only one structural ensemble is fully developed in the *Fagus sylvatica*-dominated forests of the *Fago-Quercetum* and the *Melico-Fagetum*, lower structural ensembles, often referred to as shrub layers, here obviously do regenerate in canopy gaps but never reach maturity, since they are overgrown by *F. sylvatica*.

Light Simulation

Horizontal and vertical light patterns as simulated with the program FOREYE are discussed and a first relation with inversion surfaces and vegetation pattern has been demonstrated.

The potential of the SILVI-STAR system has been discussed for different disciplines and the scope for improvements and further development has been discussed.

References

Aichinger, E., 1967. Pflanzen als forstliche Standortsanzeiger. Von Klagenfurt, Wien

Akça, A., 1983. Aerophotogrammetrische Messung der Baumkrone. Allg Forst Z 30:772-773

Akker, A., Groeneveld, W., 1984. Reconnaissance of Amazonian rainforest architecture and insect habitats. AUW Wageningen-Silviculture report 84/19

Allen, T.F.H., Starr, T.B., 1982. Hierarchy: Perspectives for ecological complexity. University of Chicago Press, Chicago

Alvey, N. et al., 1982. An introduction to Genstat. Academic Press, London

Ammon, W., 1937. Das Plenterprinzip in der Schweizerischen Forstwirtschaft. Bern Leipzig

Anderson, M.C., 1963. Light measurements in woodlands and other shaded sites. Ph. D. Thesis University of Cambridge

Anderson, M.C., 1964. Studies of the woodland light climate. J Ecol I and II 52:27-41 and 643-663

Anonymous, 1958. Study of tropical vegetation. Proc. Kandy Symp. 1956. UNESCO Paris

Anonymous, 1985. Meerjarenplan Bosbouw Beleidsvoornemen Tweede Kamer vergaderjaar 1984-1985. Nr. 18.630:1-2

Armson, K.A., Fessenden, R.J., 1973. Forest windthrows and their influence on soil morphology. Soil Science Soc. Amer. Proc. 37:781-783

Attiwill, P.M., 1962. Estimating branch dry weight and leaf area from measurements of branch girth in *Eucalyptus*. Forest Science 8:132-141

Aubréville, A., 1965. Principes d'une systématique des formations végétales tropicales. Adansonia N.S. 5:153-196

August, P.V., 1983. The role of habitat complexity and heterogeneity in structuring tropical mammal comunities. Ecology 46(6):1495-1507

Bachacou, J., Chessel, D., 1979. Étude des structures spatiales en forêt alluviale rèhnane, III Dispersion interspécifique et analyse des correspondances. Oecologia Plantarum 14(3):371-388

Baren, B. van, Hilgen, P., 1984. Structuur en dynamiek in la Tillaie een ongestoord beukenbos in het bosgebied van Fontainebleau. MSc Thesis AUW-Silviculture-RIN Leersum

Barillon d'Amoncourt, 1664. Grand livre de réformation. Archives Forestières de Fontainebleau

Barkman, J.J., 1968. Das synsystematische Problem der Mikrogesellschaften innerhalb der Biozönosen. In: Tüxen, R. (ed.) Ber. Int. Symp. Vegetationskunde Stolzenau/Weser 1964:21-53

Barkman, J.J., 1970. Enige nieuwe aspecten inzake het probleem van synusiae en microgezelschappen. AUW Misc. Papers 5:85-116

Barkman, J.J., 1973. Synusial approaches to classification. In: Whittaker(ed) Handbook of vegetation science. Junk, The Hague:437-491

Barkman, J.J., 1977. Die Erforschung des Mikroklimas in der Vegetation, theoretische und methodische Aspekte. Mededelingen No 175 Biologisch station Wijster

Barkman, J.J., 1979. The investigation of vegetation texture and structure In: Werger, J. (ed.) The study of vegetation. Junk, the Hague:125-160

Bartelink, H.H., 1988. Gebruik van luchtfoto's bij metingen aan kronen in permanente proefplots. MSc thesis Vakgroep Landmeetkunde en Teledetectie, AUW-Silviculture Wageningen

Barthelemy, J.P., 1979. Etude des structures spatiales en forêt alluviale rhènane, IV Comparison des cartes de répartition des éspeces. Oecologia Plantarum 14(3):389-399
Beatty, S.W., 1981. The role of treefalls and forest microtopography in pattern formation in understory communities. Ph. D. Thesis Cornell University
Beek, J, van der, Maessen, P.P.Th., 1981. The equipment for measuring width of annual growth rings. Ned Bosb Tijd 53 (6):158-164
Beekman, F., 1981. La dynamique d'une forêt alluviale et le role des lianes. MSc Thesis AUW-Silviculture Wageningen
Berkel, C.J.M. van, 1979. Onderzoek naar kwantitatieve methoden voor het beschrijven van vegetatiestructuren. AUW-RIN Leersum
Bijlsma, R.J., 1989. RIN Bosecologisch Informatiesysteem Documentatie applicatieprogramatuur I FOREYE CANOPIX en subprogramma's. RIN Intern Rapport
Blomqvist, A., 1879. Eine neue Methode der Holzwuchs und die Standortsvegetation bildlich darzustellen. Bidrag till kannedom af Finlans natur och folk utgifina af Finska Vetenskaps Societeten Helsingfors
Blondel, J. Cuvillier, R., 1977. Une méthode simple et rapide pour décrire les habitats d'oiseaux: le stratiscope. Oikos 29:326-331
Bobbink, R., Willems, J.H., 1987. Increasing dominance of *Brachypodium pinatum* (L) in chalk grasslands: A treath to species-rich ecosystems. Biological Conservation 40(4):301-314
Bodegom, A.J., van, 1981. Gevolgen van selectieve exploitatie in het mesofitisch regenwoud van Sarwa Suriname: de eerste 14 jaar. MSc Thesis AUW-Silviculture Wageningen
Bonhomme, R., 1976. Détermination des profils d'indice foliaire et de rayonnement dans un couvert végétal a l'aide des photographies hémisphériques. Annales Agronomiques 27(1):33-59
Bonhomme, R., Chartier, P., 1971. The interpretation and automatic measurement of hemispherical photographs to obtain sunlit foliage area and gap frequency. Isr J Agric Res 22(2):53-61
Boojh, R., Ramakrishnan, P.S., 1982. Growth strategy of trees related to successional status I. Architecture and extension growth. For Ecol and Man 4:359-374
Bormann, F.H., Likens, G.E., 1979. Pattern and process in a forested ecosystem. Springer, Berlin Heidelberg New York
Bouten, W., Water, E. de, 1987. Microgolftransmissie, nieuw gereedschap in boshydrologisch onderzoek. Rapport No 33 Fysisch Geografisch Bodemkundig Laboratorium Universiteit van Amsterdam
Bouten, W., Vré, F.M. de, Verstraten, J.M., Duysings, J.J.H.M., 1984. Carbon dioxide in the soil atmophere: Simulation model parameter estimation from field measurements. Proc. Uppsala Symp IAHS-AISH publication 150:23-30
Bouten, W., Schaap, M, Hakkaart, P, 1989. Monitoring and modelling rainfall interception and canopy wetness. Proc. Workshop Monitoring air pollution and forest ecosystem research, Bilthoven Febr. 1989
Bos, J.M. van den, 1985. Structuuronderzoek in natuurbos van Bialowieza. MSc thesis AUW-Silviculture Wageningen-RIN Leersum
Bosch, A.L., 1987. Het boservaat als onderzoeksobject. Bosbouwvoorlichting 29(1):3-5
Botkin, D.B., Janak, J.F., Wallis, J.R., 1972. Some ecological consequences of a computer model of forest growth. J Ecol 60:849-872
Bouchon, J., Faille, A., Lemée, G., Robin, A.M., Schmitt, A., 1973. Cartes et notice des sols du peuplement forestier et des groupements végéteaux de la réserve de la Tillaie (forêt de Fontainebleau). Université de Paris XI Centre d'Orsay

Brünig, E.F., 1977. Ökologische Stabilität von forstlichen Monokulturen als Problem der Bestandesstruktur. Verhandlungen der Gesellschaft fur Oekologie Göttingen. Junk The Hague:189-204

Brünig, E.F., 1981. Naturwald als Leitbild für den Waldbau. Allg Forst Z 17:413-415

Braun-Blanquet, J., 1964. Pflanzensoziologie. Wien 3 Auflage

Breemen, N., van, Burrough, P.A., Velthorst, E.J., Van Dobben, H.F., De Wit, T., Ridder, T.B., Reynders, H.F.R., 1982. Soil acidification from atmosheric ammonium sulphate in forest canopy throughfall. Nature 299(7):548-550

Brewer, R., 1980. A half-century of changes in the herb layer of a climax deciduous forest in Michigan. J Ecol 68(3):823-832

Brewer, R., Merritt, P.G., 1978. Wind throw and tree replacement in a climax beech-maple forest. Oikos 30:149-152

Brink, G., van de, Werf, S., van der, 1977. Vegetatie In: Houte de Lange, S.M., Rapport van het Veluwe onderzoek PUDOC Wageningen:37-47

Brown, J.K., 1974. Handbook for inventorying downed woody material. USDA Forestry Service General Technical Report INT-16 Intermountain Forest & Range Experimental Station Ogden Utah

Brown, J.L., 1976. Étude de la perturbation des horizontes de sol par un arbre qui se renverse et de son impact sur la pédogénèse. Can J Soil Sci 57:173-186

Bücking, W., 1986. Study of vegetation changes in natural reserves in south-west Germany. In: Fanta, J., Forestry dynamics and research in Western and Central Europe. Proc IUFRO Workshop SI.01_00 Ecosystems 1985 :231-241

Bücking, W., König, E., Krieglsteiner, G.J., Rodi, D., Schröter, H.J., Stummer, G., Wahl, U., 1982. Der Bannwald Steinhüsle. Mitteilungen der Forstlichen Versuchs- und Forschungsanstalt Baden-Württemberg Band 2

Bücking, W., Dettner, K., Grossmann, A., Haas, H., Hegelau, W., Kätzler, W., Kost, G., Mattheis, W., 1985. Der Bannwald Waldmoor-Torfstich. Mitteilungen der Forstlichen Versuchs- und Forschungsanstalt Baden-Württemberg

Bücking, W., Kätzler, W., Lange, E., Reinhardt, W., Weishaar, H., 1986. Methods for documenting succession in forest as developed and applied in natural forest reserves in south-west Germany. In: Fanta, J., Forest dynamics research in Western and Central Europe. Proc IUFRO Workshop SI.01_00 Ecosystems Wageningen 1985:265-273

Buis, J., 1985. Historia forestis Nedelandse bosgeschiedenis I en II. HES, Utrecht

Cannell, M.G.R., 1983. In: Huxley (ed.) Plant research and agroforestry Nairobi ICRAF

Carbiener, R., Schnitzler, A., 1987. L'évolution longitudinale des charactéristiques hydrologiques et phytosociologiques des forêts alluviales rhènanes de la plaine, Alsaciene, base scientifiques de la constitution de réserves naturelles chelonnées. Coll phytosociol 15:605-634

Carbiener, R., Schnitzler, A., Walter, J.M., 1988. Problèmes de dynamique forestière et de définition des stations en milieu alluvial. Coll phytosociol, Phytosociologie et Forestière. Cramer, Vaduz 15:656-686

Cheviron, D., 1985. Essai d'introduction de méthodes numeriques dans l'étude de la structure de groupements forestières. Application a deux types de forêts alluviales en plaine d'Alsace. Thèse Université Claude Bernard Lyon I

Christensen, N.L., 1977. Changes in structure, pattern and diversity associated with climax forest maturation in Piedmont North Carolina. Amer Mid Nat 97:176-188

Cobble, M.H., 1971. The shape of plant stems. Amer Mid Nat, 86:371-378

Cochet, P., 1977. Etude et culture de la forêt. Nancy école Nationale du GREF

Coo, H. de, 1981. Ontstaan en ontwikkeling van natuurlijke wilgenbossen in de Lepelaarsplassen. Agricultural University-Rijksdienst Ijsselmeerpolders Lelystad/Wageningen

Dale, V.H., Doyle, T.W., Shugart, H.H., 1985. A comparison of tree growth models. Ecol Model 29:145-169

Dam, D. van, Dobben, H.F., Ter Braak, C.J.F., Wit, T. de, 1986. Air pollution as a possible cause for the decline of some phanerogamic species in The Netherlands. Vegetatio 65:47-52

Dankers, G., 1983. De invloed van bosstructuur op het voorkomen van muizen. Huid & Haar 2:128-133

Davidson, S.E., Forman, R.T.T., 1982. Herb and shrub dynamics in a mature oak forest a thirty-year study. Bull. Torrey Bot Club 109(1):64-73

Davis, T.A.W., Richards, P.W., 1933. The vegetation of Moraballi Creek, British Guiana: an ecological study of a limited area of tropical rain forest Part I. J Ecol 21:350-384

Delcourt, H.R., Delcourt, P.A., Webb, T., 1983. Dynamic plant ecology: The spectrum of vegetation change in space and time. Quart. Sci. Rev. 1:153-175

Derkman, G., Koop, H., 1977. Structuur en verjonging van een oerbos Puszcza Bialowieza (Oost Polen). Praktijk verslag LH/Nb 1976-1977 projectnummer 70-71 Wageningen

Deuser, M., Kaufman, D.W., 1978. Habitat distribution of Peromyscus leucopus within prairie woods. Trans Kansas Aca Sci (USA) 80(3/4):151-154

Deuser, R.D., Shugart, H.H., 1978. Microhabitats in a forest-floor small-mammal fauna. Ecology 59(1):89-98

Deuser, R.D., Shugart, H.H., 1979. Niche pattern in a forest-floor small-mammal fauna. Ecology 60(1):108-118

Dierschke, H., 1978. Vegetationsentwicklung auf Kahlslägen verschiedener Laubwälder bei Göttingen I Dauerflächen-Untersuchungen 1971-1977. Phytocoenosis 7(1/4):29-42

Dijkstra,K., Paasman, J., Pereboom, F., Postema, J., 1986. Uit de geschiedenis van de Brederwiede. Ysselacademie Kampen

Dirkse, G.M., 1987. De natuur van het Nederlandse bos. Rijksinstituut voor Natuurbeheer RIN-rapport, 87/28

Dirkse, G.M., Thalen, D.C.P., 1987. De natuurfunctie van het nederlandse bos, enkele resultaten van de Vierde Bosstatistiek. Ned Bosb Tijd 58(4):116-127

Djawadi, K., 1977. Eine photogrammetrische Methode zu digitalen Erfassung und Charakterisierung der Oberflächengestalt von Waldbeständen. Dissertation Freiburg

Dogniaux, R., 1975. Variations géographiques climatiques des expositions énergetiques solaires sur des surfaces réceptrices horizontales et verticales. Inst Roy Mét Belg. Miscellanea Serie B 38

Doing-Kraft, H., 1954. L'analyse des carrés permanents. Acta Bot Neerl 3:421-424

Drent, W., Oosterbaan, M., 1984. Bossen en Begrazing; Een onderzoek naar de invloed van begrazing op de bosstructuur en bosontwikkeling in het New Forest. MSc thesis Department of Plant Eocology University Groningen, RIN Leersum

Ducrey, M., 1975. Utilisation des photographies hémisphérique pour le calcul de la perméabilite des couverts au rayonnements solaires I et II. Ann Sci For 32: 73-92 and 205-221

Duvigneaud, P. (ed.), 1971. Productivity of forest ecosystems. Proceedings UNESCO-IBP Brussels Symp 27-31 oktober 1969

Eber, W., 1972. On the light climate in forests near Göttingen and its influence on the ground flora. Scripta Geobotanica 3 Universität Göttingen

Edelin, C., 1977. Images de l'architecture des Conifères. Thèse Université Languedoc Montpellier

Edelin, C., 1984. L'architecture monopodiale: L'exemple de quelques arbres d'Asie tropicale. Thèse d'Etat USTL Montpellier

Ellenberg, H., 1939. Über Zusammensetzung, Standort und Stoffproduktion bodenfeuchter Eichen- und Buchen-Mischwaldgesellschaften Nordwestdeutschlands. Ebenda 5:3-135
Ellenberg, H., 1974. Zeigerwerte der Gefässpflanzen Mitteleuropas. Scripta Geobotanica IX Göttingen
Ellenberg, H., 1978. Vegetation Mitteleuropas mit den Alpen in ökologischer Sicht. Ulmer, Stuttgart
Erteld von, W., 1979. über die Beziehungen zwischen Schaftdurchmesser und Kronenschirmflche in Kiefernbestände. Forstwiss Zlbl 98:270-277
ERDAS, 1988. ERDAS-PC and PC Kit image processing user's guide version 7.3. ERDAS inc. Atlanta USA
ESRI, 1987. ARC/INFO users guide version 4.0. Environmental Systems Research Institute California USA
Escherich, G., 1927. Im Urwald. Georg Stilke, Berlin
Evans, G.C., Coombe, D.E., 1959. Hemispherical and woodland photography and the light climate. J Ecol 47:103-113
Faber, P.J., 1983. Concurrentie en groei van de bomen binnen een opstand. Uitvoerig Verslag Band 18(1). Rijksinstituut voor onderzoek in de bos- en landschapsbouw 'De Dorschkamp' Wageningen
Faille, A, Lemée, G., Pontailler, J.Y., 1984a. Dynamique de clarières d'une forêt inexploite (réserves biologiques Fontainebleau) I origine et état actuel des ouvertures. Acta Oecologica/ Oecologia Generalis 5(1):35-51
Faille, A., Lemée, G., Pontailler, J.Y., 1984b. Dynamique des clarires d'une forêt inexploite (réserves biologiques Fontainebleau) II Fermature des clairières actuelles. Acta Oecologica/Oecologia Generalis 5(2):181-199
Falinski, J.B., 1966. Plant communities of the international biological program study area in the Bialowieza National Park. Materialy Zakladu Fitosocjologii Stosowanej U.W. Warszawa-Bialowieza 14:1-23
Falinski, J.B., 1976. Windwürfe als Faktor der Differenzierung und der Veränderung des Urwaldbiotopes im Licht der Forschungen auf Dauerflächen. Phytocoenosis 5(2):85-108
Falinski, J.B., 1977. Research on vegetation and plant population dynamics conducted by the Bialowieza Geobotanical Station of the Warsaw University. Phytocoenosis 6(1/2):1-132
Falinski, J.B., 1986. Vegetation dynamics in temperate lowland primeval forest. Ecological studies in Bialowieza forest. Geobotany 8. Junk, Dordrecht Boston Lancaster
Falinski, J.B., 1988. Succession, regeneration and fluctuations in the Bialowieza Forest (NE Poland). Vegetatio 77:115-128
Falinski, J.B., Olszewski, J.L., 1977. Die Verteilung der Schneedekke in Mischwäldern des Bialowieza-Urwaldes. In: Dierschke, H. (ed) Vegetation und Klima:163-182
Faluere de la, 1716. Grand livre de réformation de 1716. Archives forestières Fontainebleau
Fisher, J.B., 1984. Tree architecture: relationships between structure and function. In: White, R.A., Dickison, W.C. (eds.), Contemporary problems in plant anatomy. Academic Press, Orlando Fla.:541-589
Flower, N., 1980. The management history and structure of enclosed woods in the New Forest, Hampshire. J Biogeog 7:311-328
Ford, E.D., 1976. The canopy of a Scots pine forest: description of a surface of complex roughness. Agr Met 17:9-32
Forman, R.T.T., Godron, M., 1981. Patches and structural components for landscape ecology. Bioscience 31:733-740

Franklin, J., Strahler, A. L., 1988. Invertible canopy reflectance modeling of vegetation structure in semiarid woodland. IEEE Transactions on geoscience and remote sensing 26-6:809-825
Fritts, H.C., 1976. Tree rings and climate. Academic Press, London
Fröhlich, I., 1930. Der Sudosteuropäische Urwald und seine Überführung in Wirtschaftswald. Zlbl Forstw
Geerdes, B., 1985. Ontwikkeling van de ondergroei in dennenbossen van de Sysselt bij Ede na 25 jaar. MSc thesis RU Utrecht-RIN Leersum
Géhu, J.M., 1973. Unités taxonomiques et le végétation potentielle naturelle du nord de la France. Documents Phytosociologiques 4:1-22
Geiger, R., 1961. Das Klima der bodennahen Luftschicht. Vieweg Braunsweig
Genssler, H., 1979. Naturwaldzellen in Nordrhein-Westfalen. Natur und Landschaftskunde Westfalen 15(4):97-106
Glavac, V., 1972. Zur Planung von geobotanischen Dauerbeobachtungsflächen in Waldschutzgebieten. Natur und Landschaft 47(5):139-143
Godron, M., 1971. Essai sur une approche probabiliste de l'écologie des végétaux. Thèse Montpellier C.N.R.S.No AO 2820
Goel, N. S., Grier, T., 1988. Estimation of canopy parameters for inhomogeneous vegetation canopies from reflectance data: III. TRIM: A model for radiative transfer in heterogeneous three-dimensional canopies. Remote Sensing of Environment 25:255-293
Graaf, N.R. de, 1986. A silvicultural system for natural regeneration of tropical rain forest in Suriname. AUW Wageningen
Grand-Mesnil, M.N., 1982. A propos des réserves biologiques: questions d'histoire. La Voix de la Forêt Bull. Assoc. Amis Forêt Fontainebleau 182 (1)
Greis, I., Kellomäki, S., 1981. Crown structure and stem growth of norway spruce undergrowth under varying shading. Silva Fennica 15(3):306-322
Gruber, J., Voit, 1917. Bialowies in deutscher Verwaltung. Parey, Berlin
Guillet, B., Robin, A.M., 1972. Interprétation et datation par le C^{14} d'horizons Bh de deux podzole humo-ferrugineux l'un forme sous callune, l'autre sous chenaie-hêtraie. C.R. Acad. Sci. (274):2859-2862
Hallé, F., Oldeman, R.A.A., 1970. Essai sur l'architecture et la dynamique de croissance des arbres tropicaux. Masson and Co, Paris
Hallé, F., Oldeman, R.A.A., Tomlinson, P.B., 1978. Tropical trees and forests, an architectural analysis. Springer, Berlin Heidelberg New York
Hari, P., Kaipiainen, L., Korpilahti, E., Mäkelä, A., Nilson, T., Oker-Blom, P., Ross, J., Salminen, R., 1985. Structure, radiation and photosynthetic production in coniferous stands. University of Helsinki Department of Silviculture Research Notes no:54
Heil, G.W., Diemont, W.H., 1983. Raised nutrient levels change heathland into grassland. Vegetatio 53:113-120
Henninger, J., 1983. Zeitreihen der Bestandesentwicklung in naturnahen Waldbeständen aus Luftbilder. Dissertation Universität Freiburg
Henry, J.D., Swan, J.M.A., 1974. Reconstructing forest history from live and dead plant material an approach to the study of forest succession in S-W New Hampshire. Ecology 55(4):772-783
Hermy, M., 1985. Ecolgie en fytosociologie van oude en jonge bossen in Binnen-Vlaanderen. Thesis R.U.G. Gent
Hermy, M., Stieperaere, H., 1981. An indirect gradient analysis of the ecological relationship between ancient and recent riverine woodlands to the south of Bruges. Vegetatio 44:43-49
Hesmer, H., 1934. Naturwaldzellen. Der Deutsche Forstwirt 16(13/14)

Hesmer, H., Schröder, F.G., 1963. Waldzusammensetzung und Waldbehandlung im Niedersächsischen Tiefland westlich der Weser und der Münsterischen Bucht bis Ende des 18 Jahrhunderts. Forstarchiv 34(7-8):177-208

Hill, M.O., 1979a. DECORANA: a FORTRAN program for detrended correpondance analyses and reciprocal averaging. Ecology and Systematics Cornell University New York

Hill, M.O., 1979b. TWINSPAN- A FORTRAN computer program for arranging multivariate data in an ordered two-way table by classification of individuals and attributes. Ecology and Systematics Cornell University New York

Hill, M.O., Gauch, H.G., 1980. Detrended correspondance analyses, an improved ordination technique. Vegetatio 42:47-58

Hillgarter, F.W., 1971. Waldbauliche und ertragskundliche Untersuchungen im subalpinen Fichtenurwald Scatle/Brigels. Dissertation ETH Zürich

Hillgarter, F.W., 1976. Beitrag zur Methodik der Erfassung und Beschreibung von Urwaldphasen. Proceedings IUFRO World-Congres Div.I Group I Oslo :79-83

Hoekman, D.H., 1985. Radar backscattering of forest stands. Intern. Journal Remote Sensing 6(2):325-343

Hoekman, D.H., 1987. Multiband scatterometer data analysis of forests. Intern. Journal Remote Sensing 8:1695-1707

Honda, H., 1971. Description of the form of trees by the parameters of the tree-like body: effects of the branching angle and the branch length. J Theor Biol 31:331-338

Hoogesteger, J., 1983. De natuurlijke wilgenbossen in Zuidelijk Flevoland. In: Groenhuis, B. (ed), 50 Jaar bosbouw en bosbouwkundig onderzoek in de Ijselmeerpolders. Flevobericht 216 Ministerie Verkeer & Waterstaat-RIJP

Horn, H.S., 1971. The adaptive geometry of trees. Princeton University Press

Houllier, F., 1985. Inventaires successifs en forêt: intérêt théorique et limites pratiques de l'échatillonnage partiellement renouvelé. Ann. Sci. For. 42(3):245-264

Houtzagers, G., 1954. Houtteelt der gematigde luchtstreek I. Tjeenk Willink Zwolle

Husch, B., 1971. Planning a forest inventory. Forest resources division Forestry Dept. F.A.O. Rome

Hutnik, J.R., 1952. Reproduction on windfalls in a northern hardwood stand. J For 50:693-694

Ilvessalo, Y., 1950. On the correlation between crown diameter and the stem of trees. Comm. For. Fenica 2(38):1-26

Jacquiot, C., Robin, A.M., Bedeneau, M., 1973. Réconstruction d'un ancien peuplement forestier en forêt de Fontainebleau par l'étude anatomique de charbons de bois et leur datation au C^{14}. Bulletin Soc. Bot. Fr.:120

Jahn, G., 1980. Diskussionsbeitrag zur Methode der Erfassung der Bodenvegetation auf Dauerbeobachtungsflächen in Naturwaldreservaten. Natur und Landschaft 55(4):149-150

Jahn, G., 1982. Application of vegetation science to forestry. Handbook of vegetation science XII. Junk, The Hague Boston London

James, F.C., Shugart, H.H., 1970. A quantitative method of habitat description. Audubon Field Notes 24(6):727-736

James, F.C., Wamer, N.O., 1982. Relationship between temperate forest bird communities and vegetation structure. Ecology 63(1):159-171

Jarvis, P.G., James, G.B., Landsberg, J.J., 1976. Coniferous forest. In: Monteith, J.L. (ed) Vegetation and the atmosphere Vol. II:171-240

Jonge, G., Dienske, H., 1979. Habitat and interspecific displacement of small mammals in the Netherlands. Netherlands J Zool 29(2):177-214

Jongman, R.H.G., Ter Braak, C.J.F., Tongeren, O.F.R. van, 1987. Data analysis in community and landscape ecology Pudoc Wageningen

Kalkhoven, J.T.R. et al., 1976. Landelijke milieukartering een landschapsecologische kartering van het natuurlijk milieu in Nederland t.b.v. de ruimtelijke planning op nationaal niveau. Staatsuitgeverij Den Haag

Kalkhoven, J.T.R., Van der Werf, S., 1988. Mapping the Potential Natural Vegetation. In: Küchler, A.W., I.S. Zonneveld, Vegetation mapping. Handbook of vegetation science Vol 10. Kluwer, Dordrecht:375-386

Kam, M. de, Dam, B.C. van, 1987. Scheutsterfte en bastnecrose veroorzaakt door *Sphaeropsis sapinea* in Nederland. Ned Bosb Tijd 59(6), :215-219

Karr, J.R., Freemark, K.E., 1983. Habitat selection and environmental gradients: dynamics in the stable tropics. Ecology 64(6):1481-1494

Karr, J.R., Roth, R.R., 1971. Vegetation structure and avian diversity in several New World areas. Amer Nat 105:423-435

Karvec, G., 1903. Belovezskaja Pusza. Petersburg

Kätzler, W., 1984. Zur forstlichen Aufnahme der Bannwälder in Baden-Württemberg. Mitteilungen der Forstlichen Versuchs und Foschungs Anstalt Baden-Württemberg 108:123-130

Kauppi, P., Hari, P., Kellomäki, S., 1978. A discrete time model for succession of ground cover communities after clear cutting. Oikos 30, :100-105

Kenneweg, H., 1972. Die Verwendung von Farb und Infrarot Farb Luftbildern für Zwecke der Forstlichen Photo Interpretation unter besonderer Berüksichtigung der Erkennung und Abgrenzung von Kronenschäden in Fichtenbeständen. Dissertation Universität Freiburg

Kenneweg, H., 1983. Möglichkeiten und Grenzen der Fernerkennung zur Inventur von Immisionsschäden. Allg Forst Z 30:761

Klika, J., 1940. Die Planzengesellschaften des *Alnion*-Verbandes. Preslia Vestnik Ces. Botanicke spolecnosti v Praze 28-29:16-31

Knapp, R., 1974. Cyclic successions and ecosystem approaches in vegetation dynamics. Handbook of vegetation science Part VIII. Vegetation dynamics. Junk, The Hague:93-99

Knapp, R., 1984. Sampling methods and taxon analyses in vegetation science. Junk, The Hague

Kneitz, G., 1980. Möglichkeiten der Erfassung der Fauna in Naturwaldreservaten. Natur und Landschaft 55(4):156-158

K.N.M.I., 1986. Frequentietabellen 1961/'80 Globale straling uurlijks. Publikatienr. 150-18; Frequentietabellen 1956/'80 Duur zonneschijn alle stations. Publikatienr. 150-19

Knüchel, H., 1944. ber Bestandskarten und Bestandsprofile. Schweiz Z Forstw

Koch, H., Gaisberg, E. von, 1938. Die standörtlichen und forstlichen Verhältnisse im Naturschutzgebiet Untereck. Veröffentlichung. Württembergischer Landesstelle Naturschutz 15:35-40

Koeman, C., 1963. Handleiding voor de studie van topografische kaarten van Nederland 1750-1850. Wolters Groningen

Koop, H., 1981. Vegetatiestructuur en dynamiek van twee natuurlijke bossen: het Neuenburger en Hasbrucher Urwald. Verslagen van landbouwkundige onderzoekingen 904 PUDOC Wageningen

Koop, H., 1982. Waldverjüngung, Sukzessionsmosaik und kleinstandörtliche Differenzierung infolge spontaner Waldentwicklung. Berichte Internationales Symposium Struktur und Dynamik von Wäldern Rinteln :235-273

Koop, H., 1984. Ecological monitoring of natural and semi-natural forests. In: Annual Report 1983 RIN Arnhem Leersum Texel:16-19

Koop, H., 1986a. Omvormingsbeheer naar natuurlijk bos: een paradox ?. Ned Bosb Tijd 58(1/2):2-11

Koop, H., 1986b. Ecological monitoring of natural and semi-natural forests. In: Fanta, J., 1986. Forest dynamics research in Western and Central Europe, Proc IUFRO Workshop SI.01_00 Ecosystems Wageningen:257-262

Koop, H., 1987. Vegetative reproduction of trees in some natural forest in Europe. Vegetatio 72:103-110

Koop, H., 1988. Ecological monitoring of natural and semi-natural forest. Annual Report RIN 1987 Arnhem Leersum Texel

Koop, H., Hilgen, P., 1987. Forest dynamics and regeneration mosaic shifts in unexploited beech (*Fagus sylvatica*) stands at Fontainebleau (France). For Ecol Man 20:135-150

Kopinga, J., 1979. De aanwasboor als hulpmiddel bij het opsporen van verborgen gebreken van loofbomen. Bericht 96 Rijksinstituut voor onderzoek in de Bos- en Landschapsbouw 'De Dorschkamp' Wageningen

Korpel, S., 1974. Der Badiner Urwald. Zbornik Vedeckych Prac 27(16):7-37

Korpel, S., 1982. Degree of equilibrium and dynamical changes of the forest on example of natural forests of Slovakia. Acta facultatis forestalis 24, :9-31

Korpel, S., 1987. Dynamics of the structure and development of natural beech forests in Slovakia (summary in German and Russian). Acta facultatis forestalis Zvolen-Czechoslovakia 29:59-85

Kort, L.P.F.A.M., 1989. Reconstructie van dynamiek in een beukenbos-jaarringanalyses in het bosreservaat van Fontainebleau (Fr). LUW Wageningen Silviculture-RIN

Köstler, J.N., 1953. Bildliche Darstellung des Bestockungsgefüges. Allg Forst Jagdz 125:69-76

Krause, A., Schröder, L., 1979. Vegetationskarte der Bundesrepublik Deutschland 1:200.000 Potentielle Natürliche Vegetation, Blatt CC 3118 Hamburg-West. Schriftenreihe für Vegetationskunde Bonn-Bad Godesberg 14

Kuhl, W.F., 1984. Förster sehen den Wald Rot; Erfassung der Waldschäden durch Color-Infrarot Luftbilder. LOLF Mitteilungen 9(4):23-35

Kuiper, L.C., Schoten, J.P. van, 1985. Dynamische aspecten van windworp (with Engl. Summ.). Ned Bosb Tijd 57(4):130-133

Kuiper, L.C., 1989. Stability of Douglas fir against windthrow. draft

Lacordaire, S., Lescuyer, C., 1953. Aménagement de 1853. Archives Forestières Fontainebleau

Lamprecht, H., 1969. Zur Bedeutung von Naturwald-Schutzgebieten für Landschaftspflege aus Sicht von Forstwirtschaft und Forstwissenschaft. Natur und Landschaft 44(4):87

Lamprecht, H., 1980. Zur Methodik waldkundlicher Untersuchungen in Naturwaldreservaten. Natur und Landschaft 55(4):146-147

Leemans, R., 1986. Structure of the primaeval forest of Fiby. In: Fanta, J., Forest dynamics research in Western and Central Europe. Proc IUFRO Workshop SI-01-00 Ecosystems 1985:221-230

Leibundgut, H., 1945. Waldbauliche Untersuchungen über den Aufbau von Plenterwälder. Mitteilungen Schweizerische Anstalt forstliches Versuchswesen XXIV(1)

Leibundgut, H., 1959. Über Zweck und Methodik der Struktur- und Zuwachsanalyse von Urwäldern. Schweiz Z Forstw 110(3):111-124

Leibundgut, H., 1966. Waldreservate. Schweiz Z Forstw 117(12):900-907

Leibundgut, H., 1978a. Die Waldpflege 2 Aufl. Bern, Stuttgart

Leibundgut, H., 1978b. Über die Dynamik europäischer Urwälder. Allg Forst Z 24:686-690

Leibundgut, H., 1982. Europäische Urwälder der Bergstufe. Haupt, Bern Switzerland

Lemée, G., 1966. Sur l'intérêt écologiques des réserves biologiques de la forêt de Fontainebleau. Bull. Soc. Bot. Fr. 113:305-323

Lemée, G., 1978. La hêtraie naturelle de Fontainebleau. In: Problemes d'écologie: Structure et fonctionement des écosystemes terrestres. Masson, Paris:75-128
Lemée, G., 1981. Contribution a l'histoire des landes de la forêt de Fontainebleau d'apres l'analyse pollinique des sols. Bull. Soc. Bot. Fr. 128 Lett. Bot. 3:189-200
Lemée, G., Faille, A., Pontailler, J.Y., 1986. Dynamique de cicatrisation des ouvertures dans des réserves biologiques de la forêt de Fontainebleau. In: Fanta, J., Forest dynamics research in Western and Central Europe Proc IUFRO Workshop SI.01-00 Ecosystems 1985:170-181
Lescure, J.P., 1978. An architectural study of the vegetation's regeneration in French Guiana. Vegetatio 37(1):53-60
Li, X., Strahler, A. L., 1985. Geometric-optical modeling of a conifer forest canopy. IEEE Transactions on geoscience and remote sensing GE-23-5:705-721
Li, X., Strahler, A. L., 1986. Geometric-optical bidirectional reflectance modeling of a conifer forest model. IEEE Transactions on geoscience and remote sensing GE-24-6:906-919
Linden van der, J.A., 1973. Topografische en Militaire kaart van het Koninkrijk der Nederlanden. Fibula-van Dishoeck Bussum
Londo, G., 1971. Patroon en proces in duinvalleivegetaties langs een gegraven meer in de Kennemerduinen. Dissertatie Nijmegen Verhandelingen No 2 RIN, Leersum
Londo, G., 1977. Bossen en natuurbeheer. Ned Bosb Tijd 49(7/8):219-228
Londo, G., 1978. A method of studying vegetation dynamics in a two-dimensional microgradient. Phytocoenosis 7(1-4):165-176
Londo, G., 1982. Kleinschalige natuurbouw proeftuin Scherpenzeel, Doel, milieu en methoden. RIN-rapport 82/15
Londo, G., 1984. The decimal scale for relevés of permanent quadrats. In: Knapp, R., Sampling methods and taxon analyses in vegetation science. Junk, The Hague:46-370
Loon, H. van, Mensink, A., Scheltinga, A., 1985. Vegetatiekundig onderzoek in verschillende boscomplexen in het Gerendal. MSc Thesis RIN Leersum-Department of Botany II Catholic University Nijmegen
Lustgraaf, B., van de, 1984. Remote sensing methoden voor het beoordelen van de gezondheidstoestand van vegetaties. Pudoc Wageningen
Lutz, H.J., 1940. Disturbance of forest soil resulting from uprooting of trees. Yale University School of Forestry Bulletin. No 45
Lyford, W.H., Maclean, D.W., 1966. Mound and pit microrelief in relation to soil disturbance and tree distribution in New Brunswick. Canada Harvard Forest Paper 15
Mabelis, A.A., 1977. Artenreichtum von Ameisen in einiger Waldtypen. Ber. Int. Symp. Vegetation und Fauna Rinteln:187-208
MacArthur, R.H., MacArthur, J.W., 1961. On bird species diversity. Ecology 42:594-598
Malmer, N., Lindgren, L., Persson, S., 1978. Vegetational succession in a south Swedish deciduous wood. Vegetatio 36(1):17-29
Mansour, S., 1986. Nabootsingen van boomkronen en toepassing daarvan bij simulatie van fish-eye foto's. MSc Thesis AUW-Silviculture-Infomatica-RIN Leersum
Markgraf, F., Dengler, A., 1931. Aus den Südosteuropäischen Urwäldern. Z Forst Jagdw 63(1):1-32
Martin, A.J., 1976. Suitability of the line intersect method for sampling hardwood logging residues. USDA Forest Service Research Paper NE 339
Masumi, A., 1984. Interpretationsschlüssel zur Auswertung von Infrarot Farbluftbildern für die Waldschädeninventur. Allg Forst Z 27:687-689
Maurer, H., Meier, E., Häfner, H., Frei, U., 1983. Automatisierte Baumarten-kartierung mit grossmasstäblichen Infra Rot Farbbildern. Schweiz Z Forstw 134(1):37-47

Mauve, K., 1931. Über Bestandesaufbau, Zuwachsverhältnis und Verjüngung im Galizischen Karpaten-Urwald. Hannover

Mayer, H., 1976. Waldbauliche Behandlung von Schutzwäldern. Proc XVI IUFRO World-Congress Oslo Group I:24-35

Mayer, H., 1984. Waldbau auf soziologisch-oecologischer Grundlage. Fisher, Stuttgart

Mayer, H., Neumann, M., 1981. Struktureller und entwicklungsdynamischer Vergleich der Fichten-Tannen-Buchen Urwälder Rotwald/Niederstereich und Corkova Uvala. Zlbl ges Forstw 98(1):1-29

Mayer, H., Neumann, M., Schrempf, W., 1979. Der Urwald Rotwald in den Niederösterreichischen Kalkalpen. Jahrbuch Vereinigung zum Schutze der Bergwelt e.V. München 44:79-117

Mayer-Wegelin, H., Muhring/Schulz-Bruggemann, 1952. Untersuchungen ber den Bestandesaufbau im kleinen Urwald des Rotwaldes. Zlbl Forst Holzw

McMahon, T.A., Kronauer, R.E., 1976. Tree structures: deducing the principle of mechanical design. J Theor Biol 59:443-466

Mitscherlich, G., 1971. Wald, Wachstum und Umwelt. 2 Band: Waldklima und Wasserhaushalt. J.D. Saurländer, Frankfurt/Main

Mohren, G.M.J., 1987. Simulation of forest growth, applied to Douglas fir stands in the Netherlands. Thesis AUW Wageningen

Moon, P., Spencer, D.C., 1942. Illumination from a non-uniform sky. Trans. Illum. Engng. Soc. New York 37:707-712

Mozgawa, J., 1983. Photogrammetrische Nachweis von Veränderungen der Kronenformen in immissionsgeschädigten Fichtenbeständen. Allg Forst Z 30:774-776

Mückenhausen, E., 1977a. Die Bodenkunde. DLG GmbH, Frankfurt/Main

Mückenhausen, E., 1977b. Entstehung Eigenschaften und Systematik der Böden der Bundesrepublik Deutschland. DLG GmbH, Frankfurt/Main

Mueller-Dombois, D., Ellenberg, H., 1974. Aims and methods of vegetation ecology. Wiley, New York

Mühlenberg, M., 1980. Kann man mit Hilfe von Vegetationsstruktur-Messungen Veränderungen von Tiergemeinschaften bestimmen ? Ber. Int. Symp. Epharmonie Rinteln:309-327

Nagano, M., 1978. Biological production in a warm temperate evergreen oak forest of Japan: Dynamics and stand development. In: Kira, T, Ono, Y., Hosokawa, T., I.B.P. Synthesis 18:21-32

Naka, K., 1982. Community dynamics of evergreen broadleaf forest in south-western Japan I. Wind damaged trees and canopy gaps in an oak forest. Bot Mag Tokyo 95:385-399

Nakashizuka, T., 1985. Diffused light conditions in canopy gaps in a beech (*Fagus crenata* Blume) forest. Oecologia 66:472-474

Nakashizuka, T., Numata, M., 1982. Regeneration process of climax beech forests I. Structure of a beech forest with the undergrowth of Sasa. Japanese J Ecol 32:57-67

Newbould, P.J., 1967. Methods for estimating the primary production of forests. IBP handbook nr. 2. Blackwell, Oxford

Niemann, E., 1968. Gedanken zur Problematik von Totalreservaten in Wäldern. Archif Naturschutz und Landschaftsforschung 8(4):273-290

Noirfalise, A., 1987. Vegetation map (1:3.000.000) of the Council of Europe member states. Nature and environmental series European committy for the conservation of nature and natural resources Strasbourg No 16

O'Neil, R., DeAngelis, D.L., Waide, J.B., Allan, T.F.H., 1986. A hierarchical concept of ecosystems. Princeton University Press

Ogawa, H., Yoda, K., Kira, T, Ogino, K., Shidei, T., Ratanawongse, D., Apasutaya, D., 1965. Comparative ecological study of three main types of forest vegetation in Thailand I and II. Nature and life in South East Asia 4:13-48 and 49-80

Ohsawa, M., 1981. A basic unit in forest community dynamics: A case study in the subalpine forests of Japan. Proc Symp Forest regeneration in Southeast Asia Bogor 1979 BIOTROP special publications No 13:43-62

Oker-Blom, P., 1986. Photosynthetic radiation regime and canopy structure in modeled forest stands. Acta Forestalia Fennica 197:1-44

Oldeman, R.A.A., 1972. L'architecture de la végétation ripicole forestière des fleuves et criques guyanais. Adansonia (N.S.) 12(2):253-265

Oldeman, R.A.A., 1974a. L'architecture de la forêt guyanaise. Mémoires ORSTOM. 73

Oldeman, R.A.A., 1974b. Écotopes des arbres et gradients écologiques verticaux en forêt guyanaise. La Tèrre et la Vie 28(4):487-520

Oldeman, R.A.A., 1978. Architecture and energy exchange of dicotyledonous trees in the forest. In: Tomlinson, P.B., Zimmerman, M.H. (ed.), Tropical trees as living systems Chapter 23. Cambridge University Press, London New York

Oldeman, R.A.A., 1979. Quelques aspects quantificables de l'arborigenèse et de la sylvignèse. Oecologia Plantarum 14(3):1-24

Oldeman, R.A.A., 1980. Field guide for the research group , Scale-drawing and architectural analysis of vegetations. Inst. of Ecology Padjadjaran Univ. Bandung Indonesia-Dep. of Silviculture Agricultural Univ. Wageneningen Netherlands

Oldeman, R.A.A., 1983. Tropical rain forest, architecture, sylvigenesis and diversity. In: Sutton, S.L., Whitmore, T.C., Chadwick, A.C., Tropical rain forests: ecology and management. Blackwell, Oxford:139-150

Oldeman, R.A.A., 1986. On the limits of forest ecosystems: do they have skins ?. IUFRO Ljubljana Division 6

Oldeman, R.A.A., 1989. Elements of silvology. Springer, Berlin Heidelberg New York Tokyo (in press)

Oldeman, R.A.A., Schmidt, P., 1986. Silvicultural design an exercise for students. In: Fanta, J., Forest dynamics and research in Western and Central Europe. Proceedings IUFRO workshop SI.01_00 Ecosystems 1985 :205-210

Oliver, C.D., Stephens, E.P., 1977. Reconstruction of a mixed-species forest in Central New England. Ecology 58(3):562-572

Olszewski, J.L., 1968. Role of uprooted trees in the movements of rodents in forest. Oikos 19:99-104

Olszewski, J.L., 1976. Relation between the amount of rainfall reaching the forest floor and the amount over a mixed deciduous forest. Phytocoenosis 5(2):127-156

Ondok, J.P., 1984. Simulation of stand geometry in photosynthethic models based on hemispherical photographs. Phytosynthetica 18:231-239

Opdam, P., Bladeren, G.J. van, 1981. De broedvogelbevolking van beheerde en onbeheerde delen van het Forstamt Hasbruch (Oldenburgerland, BRD) relatie tot bosstructuur. RIN Leersum report 81/2

Opdam, P., Schotman, A., 1986. Betekenis van structuur en beheer van bossen voor de vogelrijkdom. Ned Bosb Tijd 58(12):21-33

ORACLE Corporation, 1987. SQL*plus user's guide version 5.0. ORACLE corporation California USA

Paczoski, J., 1930. Lasy Bialowieza - Die Waldtypen Bialowieza. Monogafie Naukwe 1:1-575

Passarge, H., 1968. Neue Vorschläge zur Systematic nordmitteleuropaischer Waldgesellschaften. Feddes Repert 77:75-103

Persson, S., 1980. Succession in a south Swedish deciduous wood: a numerical approach. Vegetatio 43:103-122

Peterken G.F., Tubbs, C.R., 1965. Woodland regeneration in the New Forest, Hampshire since 1650. App Ecol 2 (1):59-170

Peterken, G.F., 1974. A method for assessing woodland flora for conservation using indicator species. Biol Cons 6 (4):239-241

Peterken, G.F., 1981. Woodland conservation and management. Chapman & Hall, London New York

Peterken, G.F., Backmeroff, C., 1988. Long-term monitoring in unmanaged woodland nature reserves. Research & survey in nature conservation. Nature Conservancy Council No. 9

Peterken, G.F., Jones, E.W., 1987. Forty years of change in Lady Park Wood: The old growth stands. J Ecol 75:477-512

Piepers, A., 1986. De habitatvoorkeur van kleine zoogdieren (met name muizen). MSc thesis AUW Wageningen VPO-RIN Leersum

Piotrowska, H., 1978. Zu methodischen Problemen der Sukzessionsuntersuchungen auf Dauerflächen. Phytocoenosis 7(1/4):177-189

Plaat, G.J. van der, 1984. Faunistische inbreng in het bosecologisch onderzoek; een voorstudie. RIN Arnhem report 84/5

Pontailler, J. Y., 1979. La régéneration du Hêtre en forêt de Fontainebleau, ses rélations avec les conditions hydriques stationnelles. Thèse Université de Paris-sud centre Orsay

Pott, R., 1981. Der Einfluss der Niederholzwirtschaft auf die Physiognomie und die floristisch-soziologische Struktur von Kalkbuchenwälder. Tuexenia (1):233-242

Prince, S.D., 1987. Measurement of canopy interception of solar radiation by stands of trees in sparsely wooded savanna. Int J Remote Sensing 8(12):1747-1766

Proctor, M.C.F., 1980. Estimates from hemispherical photographs of radiation climates of some bryophyte habitats in the British Isles. J Bryol 11:351-366

Prusa, E., 1985. Vegetace CSSR A15, Die Böhmischen und Mährischen Urwälder - Ihre Struktur und ökologie. Academia Praha

Putz, F.E., 1983. Treefall pits and mounds, buried seeds and the importance of soil disturbance to pioneer trees on Barro Colorado Island Panama. Ecology 64:1069-1074

Rackham, O., 1976. Trees and woodland in the British landscape. Dent & Sons Ltd, London

Rackham, O., 1980. Ancient woodlands their plants life history and use. Arnolds, London

Rauner, J.L., 1976. Deciduous forests. In: Monteith, J.L. (ed). Vegetation and the atmosphere, Case studies Vol II:241-264

Reffye, P. de, 1979. Modèlisation de l'architecture des arbres par des processus stochastiques. Simulation spatiales des modèles tropicaux sous l'effet de la pesanteur. Application au *Coffea robusta*. Thèse Université de Paris-sud centre d'Orsay

Reffye, P. de, Edelin, C., 1989. La modèlisation de la croissance des plantes. La Recherche 207:158-168

Rhody, B., 1983. Erfassung mitteleuropaischer Hauptbaumarten im Rahmen von Waldinventuren mit Hilfe kleinformatiger Luftaufnahmen. Schweiz Z Forstw 134(1):17-36

Richards, P.W., 1952. The tropical rain forest. Cambridge University Press

Richards, P.W., 1983. The three-dimensional structure of tropical rain forest. In: Sutton, S.L., Whitmore, T.C., Chadwick, A.C., Tropical rain forests: Ecology and management. Blackwell, Oxford:3-10

Rijken, M., 1976. Comparison of three maps of a natural forest vegetation obtained by different methods. Phytocoenosis 5(2):109-126

Rollet, B., 1974. L'architecture des forêts denses humides sempervirentes de plaine. Nogent-sur-Marne CTFT

Roloff, A., 1985. Morphologie der Kronenentwicklung von *Fagus sylvatica* L. (Rotbuche) unter besonderer Berücksichtigung möglicherweise neuartiger Veränderungen. Dissertation Universität Göttingen

Rook, D.A., Grace, J.C., Beets, P.N., Whitehead, D., Santantonio, D., Madgwick, H.A.I., 1985. Forest design: biological models and management implications, In: Cannel, J.F. et al. Atributes of trees as crop plants. IUFRO conference Edinburg 1985

Ross, J., 1981. The radiation regime and architecture of plant stands. Junk, The Hague

Rozendaal, J., Slootweg, R., 1982. Onderzoek naar veranderingen in Zuidlimburgse hellingbossen en mogelijkheden van principale komponentenanalyse bij beschrijving. MSc Thesis RIN-Department of Ecology University of Leiden

Satterlund, D.R., 1983. Forest shadows: How much shelter in a shelterwood ?. For Ecol Man 5:27-37

Scamoni, A., 1953. Naturwaldzellen. Natur und Heimat 6:176

Schauer, W., Stöcker, G., 1976. Bestockungsanalysen und Dauerbeobachtungsflächen in Naturschutzgebieten naturnaher Berg- und Fichtenwälder. Beitr Forstwirt 10(3):147-153

Schmidt, W., 1988. Langjährige Veränderungen der Krautschicht eines Kalkbuchenwaldes (Dauerflächenuntersuchungen). Tuexenia 8:327-338

Schneider, T., Bresser, A.H.M., 1988. Dutch priority programme on acidification. Evaluation report 00-06 National Institute of Public Health and Environmental Protection Bilthoven

Schowengerdt, R.A., 1983. Techniques for image processing and classification in remote sensing. Academic Press, New York

Schulting, R., Staaveren, C., van, 1981. Inventarisatie methoden voor bosreservaten. Thesis HBCS Velp-RIN Leersum

Schulz. J.P., 1960. The vegetation of Suriname Vol.II Ecological studies on rain forest in nothern Suriname. Verhandelingen Koninklijke Nederlandse Akademie Wetenschappen afdeling Natuurkunde 253(1):1-267

Schutz, P., Tol, G. van, 1989. Vitaliteit van het Nederlandse bos. Ned Bosb Tijd 61(2):50-51

Shugart, H.H., 1984. A theory of forest dynamics, The implications of forest succession models. Springer, New York Berlin Heidelberg Tokyo

Sillevis, J., Kraan, H., 1986. De school van Barbizon Franse meesters van de 19e eeuw. Haags gemeentemuseum

Simon, H.A., 1962. The architecture of complexity. Proc Am Philos Soc (106):467-482

Smiet, A.C..,1989. Human impact on moutain forest in the river Konto area: Vegetation and transect studies. Konto River Project, Project communications No. 11 Malang

Smittenberg, J.H., 1969. Type-inventarisatie van broekbossen in het plassengebied ten oosten van de Utrechtse Vecht. RIVON-rapport Zeist

Smittenberg, J.H., 1976. De vegetatie van het moerasboscomplex "De Suikerpot" te Kortenhoef. In: Bakker P.A. et al. (ed). De Noordelijke Vechtplassen: 181- 195. Sticht. Cie. Vecht en Oost. en West. Plassengebied Vlaardingen

Sollins, P., 1982. Input and decay of coarse woody debris in coniferous stands in Western Oregon and Washington. Can J For Res 12:18-28

Spellmann, H., von, 1984. Zustandserfassung in Kiefernbeständen mit Hilfe des Luftbildes. Dissertation Universität Göttingen

Spellmann, H., von, 1986. Dendrometrische Luftbildauswertung in Kiefernbeständen für Zwecke der Forsteinrichtung. Forstarchiv 57:20-24

Stephens, E.P., 1956. The uprooting of trees, a forest process. Soil science Proc. 20:113-116

Stichting voor Bodemkartering, 1965. De Bodem van Nederland. Wageningen

Stocki, J., 1981. Preliminary study on insect life-spaces in Oostereng forest. AUW-Silviculture Wageningen Stone, E.L., 1975. Windthrow influence on spatial heterogeneity in a forest soil. Mitteilungen der Schweizerischen Anstalt für forstliches Versuchswesen 51:77-87

Sykes, J.M., Horrill, A.D., Mountford, M.D., 1983. Use of visual cover assesments as quantitative estimators of some British woodland taxa. J Ecol 71:437-450

Tendron, G., 1983. La forêt de Fontainebleau, de écologie à la sylviculture. Office Nationales des Forêts Fontainebleau

Ter Braak, C.J.F., 1986. Canonical correspondence analyses: a new eigenvector technique for multivariate direct gradient analyses. Ecology 67(5), :1167-1179

Ter Braak, C.J.F., 1987. CANOCO a FORTRAN program for canonical comunity ordination by [partial][detrended][canonical] correspondence analyses, principal components analyses and redundancy analyses (version 2.1). TNO Institute of applied computer science Wageningen

Ter Braak, C.J.F., N.J.M. Gremmen, 1987. Ecological amplitudes of plant species and the internal consistency of Ellenberg's indicator values for moisture. Vegetatio 69:79-87

Terborgh, J., 1985. The vertical component of plant species diversity in temperate and tropical forests. Amer Nat 126 (6):760-776

Thalen, D.C.P., 1980. On photographic techniques in permanent plot studies, Vegetation Dynamics. Proc Second Symposium Working group Succession Reasearch on permanent plots Ierseke 1975 Junk:19-24

Thijsse, J.P., 1939. Een nieuw natuurmonument. De Levende Natuur 8:225-230

Thom, A.S., Stewart, J.B., Oliver, H.R., et al., 1975. Comparison of aerodynamic and energy bedget estimates of fluxes over a pine forest. Quart J Roy Met Soc 101:93-105

Tongeren, O. van, 1984. Concurentie en groei van bomen binnen een opstand, een bosbouwmodel beschouwd vanuit natuurbeheer. Vakblad voor biologen 64(8), :146-147

Tonne, F., 1954. Besser Bauen mit Besonnungs und Tageslicht Planung. Karl Hoffmann, Schondorf

Torquebiau, E.F., 1985. The tropical rain forest mosaic pattern with special reference to Indonesia. BIOTROP Bogor

Torquebiau, E.F., 1986. Tropical rain forest regeneration patterns as recognized by laying down forest mosaic maps. Proc Symp Forest regeneration in Southeast Asia Bogor 1984 BIOTROP special publications No 25:139-150

Trautmann, W., 1969. Zur Einrichtung von Naturwaldreservaten in der Bundesrepublik Deutschland. Natur und Landschaft 44(4):88-89

Trautmann, W., 1972. Potentielle natürliche Vegetation. In: Deutscher Planungsatlas Band I. Janecke, Hannover

Tubbs, C.R., 1986. The New Forest. The new naturalist. Collins, London

Tüxen, R., 1956. Die Pflanzengesellschaften Nordwestdeutschlands. Gartenbauamt Bremen

Urban, D.L., O'Neil, R.V., Shugart, H.H., 1987. Landscape ecology, A hierarchical perspective can help scientists understand spatial patterns. Bioscience 37(2):119-127

Vandeursen, J., Wisse, J., 1985. De invloed van fluviatile dynamiek op de groei en structuur van een natuurlijk *Fraxino-Ulmetum* in de Elzas. MSc Thesis RIN Leersum, AUW-Silviculture Wageningen

Veen, A.W.L., 1984. Onderschepping en verdeling van hemelwater in bossen. Ned Bosb Tijd 56(11):270-279

Veerkamp, Onderzoeks project paddestoeleninventarisatie (in prep)

Veldkamp, R., 1985. Broedvogels in de Wieden. Rapport Vereniging tot Behoud van Natuurmonumenten in Nederland

Vickery, W.L., 1981. Habitat use by Northeastern forest rodents. Amer Midland Nat 106(1):111-118

Vlieger, J., 1937. Plantensociologische aanteekeningen in de omgeving van Blokzijl. Nederlands Kruidkundig Archief 47:374-388

Vooren, A.P., Offermans, D.M.J., 1985. An ultra-light aircraft for low-cost, large-scale stereoscopic aerial photographs. Biotropica 17:84-88

Vuure, T. van, 1985. Zoogdieren, bossen en wederzijdse invloeden. PUDOC Wageningen Vyskot, M. a kolectiv, 1981. Ceskoslovenske Pralesy. Academia Praha

Walter, J.M.N., 1979. Étude de structures spatiales en forêt alluviale rhènane, I Problemes structureaux et données expirimentales. Oecologia Plantarum 14(3):345-359

Watt, A.S., 1923. On the ecology of British beech woods with special reference to their regeneration, Failure of regeneration Part I. J Ecol 11(1):1-48

Watt, A.S., 1947. Pattern and process in the plant community. J Ecol 35:1-22

Werf, S. van der, 1989. De bostypen van Nederland. PUDOC Wageningen (in press)

Werger, M.J.A., Zukrigl, K., Kley van der, A., 1984. Struktur einiger Laubwälder im Niederöstereichischen Weinviertel. Flora 175:31-44

West, D.C., Shugart, H.H., Botkin, D., 1981. Forest succession: concepts and application. Springer, New York Heidelberg Berlin Tokyo

Westhoff, V., Bakker, P.A., Leeuwen, C.G., Voo, E.E. van der, 1971. Wilde Planten, Flora en vegetatie van onze natuurgebieden deel 2: het lage land. Vereniging tot behoud van natuurmonumenten

Westhoff, V., Held, A.J. den, 1969. Plantengemeenschappen in Nederland. Thieme & Cie, Zutphen

Whitmore, T.C., 1975. Tropical rain forests of the Far East. Clarendon Press Oxford

Wiegers, J., 1985. Succession in fen woodland ecosystems in the Dutch Haf district, with special reference to *Betula pubescens* Ehr.. Dissertation University Amsterdam

Wieren, S.E. van, Borghesius, J.J., 1988. Evaluatie van bosbegrazingsobjecten in Nederland. RIN Arnhem report 88/63

Wieren, S.E., Koop, H., Beije, H.M., Thalen, D.C.P., 1989. Willson, M.F., 1974. Avian community organisation and habitat structure. Ecology 55:1017-1029

Wilmanns, O., 1977. On forest reserves in the German Federal Republic. Nat Can 104:175-180

Winckel, R. van de, 1980. Het Wyhlerwald, architectuur van een Rijnoeverbos. MSc Thesis AUW-Silviculture Wageningen

Winckel, R. van de, 1984. L'architecture et la dynamique d'une forêt alluviale rhènane sauvage. Colloques Phytosociologiques La végétation des forêts alluviales. Cramer, Vaduz 9:503-543

Wirdum, G. van, 1979. Dynamic aspects of trophic gradients in a mire complex. Committee for Hydrological Research TNO: The relation between water quantity and water quality in studies of surface waters; Proc and Information 25:66-82

Wirdum, G. van, 1987. The role of spatial variety in the application of ecology to nature protection and agriculture. Annual report RIN 1986:31-33

Wit, C.T. de et al., 1978. Simulation of assimilation, respiration and transpiration of crops. Simulation Monographs PUDOC Wageningen

Wit, Toke, 1982. Permanent plots, cryptogamic plant species and air polution. In: Steubing, L., Jager, H.J., (eds) Monotoring of air pollutants by plants. Junk, The Hague: 53-58

Wloczewski, T., 1972. Dynamika rozwoju drzewostanow w oddziale 319 Bialowieskiego Parku Narodowego. Folia Forestalia A.20

Wolf, G., 1980. Zur vegetationskundlichen Bestandeserfassung in den Naturwaldreservaten. Natur und Landschaft 55(4):148-149

Yoda, K., 1974. Three dimensional distribution of light intensity in a tropical forest of West Malaysia. Jap J Ecol 24:247-254

Yoneda, T., 1975. Studies on the rate of decay of wood litter on the forest floor I and II. Jap J For 25(1):40-46 and 25(3):132-140

Zeide, B., 1981. Method of mound dating. For Sci 27(1):39-41

Zimmermann, H.M., Brown, C.L., 1971. Trees, Structure and function. Springer, Berlin Heidelberg New York

Zinderen-Bakker, E.M. van, 1942. Successie-onderzoek in het Naardermeer I en II. De Levende natuur 47(1):1-7 and 47(2):25-29

Zukrigl, K., 1970. Zusammenhänge zwischen Standort, Vegetation und Bestandesstruktur in Urwaldbeständen verschiedener Ausbildungen des *Abieto-Fagetum*. Posebna-izdanja XV(4):47-59

Zukrigl, K., 1973. Montane und subalpine Waldgesellschaften am Alpenostrand. Mitteilungen der forstlichen Bundes Versuchsanstalt 101

Zukrigl, K., 1978. Der Rothwald als Forschungsobjekt. Allg Forst Z 24:700-702

Zukrigl, K., 1983. Naturwaldreservate in Österreich. Oeko 5(2):20-27

Zukrigl, K., Eckhart, G., Nather, J., 1963. Standortkundliche und waldbauliche Untersuchungen in Urwaldresten der Niederösterreichischen Kalkalpen. Mitteilungen der forstlichen Bundes Versuchsanstalt Mariabrunn 62 Wien

Appendix

Manpower
One scientific worker with assistant
Fieldwork and program development

Assistance
Photography department
For sequential ground photography

Reproduction
Blueprint machine for tracing sheet reproduction

System operator

Equipment

Field station:
Hand-held computer HUSKY-HUNTER (model: 352 K)
PC Computer ATARI-1040ST
Printer STAR NX-1000
Modem for a communication with the VAX-host computer

DEC VAX-750 computer
Plotter HP and CALCOMP

PC OLIVETTI-M 280 with A3-Plotter HP-7475A and Printer MANNUSMANN Tally 88
IBM compatible diskdrive for 5 inch floppy disks (for data exchange from ATARI, 3 1/2 inch micro floppy disks to OLIVETTI)

Core Area Tree Classification

Qualitative Characteristics of Live Trees

a Species
Tree species are coded according to a numerical index of the "Naturwaldreservate" program in the BRD (Lamprecht 1980).

b Vitality
Vitality class is recorded using the IUFRO guidelines (Leibundgut 1956; Mayer 1976) in a three-class scale:
- rankly developed trees (1)
- normally developed trees (2)
- indigently developed trees (3)

c Inner crown cover
Inner crown cover is estimated as the cover inside the crown projection of a

tree, using the decimal scale of Londo (1984):
- cover from 0 to 15 percent (1)
- cover from 15 to 25 percent (2)
- cover from 25 to 35 percent (3)
....
- cover from 85 to 100 percent (9)

d Growth potential
Growth potential is classified after Hall et al. (1978):
- potential tree (1)
- tree of the present (2)
- tree of the past (3)

e Damage
Cause of damage is recorded after IUFRO guidelines:
- no visible damage (1)
- abiotic damage (2)
- biotic damage other than by game (3)
- complex damage (4 combination of 2 and 3)
- damage by game (5)

This code is completed with a second code number for the rate of damage:
- no damage (0 in combination with no visible damage in 10)
- little damage (1)
- moderate damage (2)
- severe damage (3)

Qualitative Characteristics of Dead Trees

a Dead wood position
For dead trees the tree position is expressed in the same code number:
- hanging dead trees (broken or uprooted, not yet on the forest floor) (6)
- dead trees or stumps standing upright (7)
- dead trees on the forest floor (8)
- stumps of trees cut by man (stem removed) (9)

b Rate of decay of dead wood
The foregoing code is completed with a second code number for the rate of decay of dead wood:
- fresh stump, trunk or branch, wood hard, intact bark, stem round in transverse section (1)
- wood can be superficially squashed (up to 1 cm), bark loose and partly dropped, stem round in transverse section (2)
- wood soft, most parts of the stem can be squashed up to several cm, bark dropped (except for species that first moulder inside an intact bark, such as *Ilex* and *Betula*, then to be judged on hardness of the wood, stem round in transverse section (3)
- thoroughly decayed soft wood, stem oval in transverse section, stem shows big gaps, the wood lost its consistency and disintegrates when touched (4)
- remains of the tree can only be recognized in the litter layer or are marked by a deviant vegetation (5)
- stumps of naturally died or uprooted trees that have been cut and removed by man afterwards (9). For example, an uprooting with a cut stump and a removed stem receives code 89 (removed lying dead wood). The stump of cut, already dead, trees (code 79) mostly cannot be distinguished from the one of cut living trees (code 72 for example). If dead trees attacked by the dutch elm

disease are cut to prevent the spread of the disease by bark beatles, such trees may be coded with 79.
- A code is added for trees or logs that have disappeared since a previous observation (0). The cause of their loss may be mouldering, drift (in flooded forests), or harvest by man.

Parameters of Uprootings

a Depth of the pit and height of the mound of uprootings
Depth of the pit and height of the mound of the micro relief due to uprooting are measured at the deepest and highest points respectively, using the forest floor surface as the zero level for perpendicular measuring lines.

b Rate of flattening of micro-relief due to uprooting
The rate of flattening of micro-relief due to uprooting is classified according to Koop (1981):
- recently uprooted tree, little of the soil clinging to the root system dropped (1)
- the mass of earth above the main root zone has fallen, roots and stem are partly decayed, the pit is partly filled up with litter, mixed with the fallen earth mass from the root system (2)
- the tree has mouldered away and the pit is filled up for the greater part (3)
- height of the mound decreased, the slight micro-relief is hardly recognizable (4)

c Charting of uprootings
Apart from standing and fallen trees, the mound and pit of uprootings are charted.

Description of Computer Programs

Scheme of Processing of Vegetation Data

TRANSASS.HBA
Husky BASIC program for the assessment of the herbaceous transect area plots of 2 x 2 m^2. A species list of a previous assessment is provided on the computer screen. A search routine enables to find a species by its abbrevation. New species can be added and species names can be edited during the program run. The sum of species cover of individual species is checked against the total cover. Corrections can be made at any time in any of the relevés.

VEGINPUT
VAX-FORTRAN-77 for input from field form in files (*.FUL) of vegetation data.

CONVERS
GFA-BASIC program on the ATARI for conversion of the format of data files from the HUSKI to *.VEG and *.OPN files, containing respectively species cover data and data on total cover of vegetation layers per relevé,that can be loaded with the program VEGINORA in the ORACLE database.

OPNACON
VAX-FORTRAN-77 program for conversion of vegetation data files from the editor (*.DAT) or from the program VEGINPUT (*.FUL) to *.VEG and *.OPN files, containing respectively species cover data and data on total cover of vegetation layers data of per relevé,that can be loaded with the program VEGINORA in the ORACLE database.

VEGINORA This DCL procedure generates a SQL command procedure that loads vegetation data files (*.DAT or *.FUL) in the ORACLE database.

ORACLEX
This VAX/VMS DCL command procedure uses SQL commands and selects and sorts vegetation data from the ORACLE database on year, location of the core area, species or vegetation layer. A standard file (VEGETATION.LIS) with selected tree data is the output of the program.

DECIM
A VAX-FORTRAN-77 program for conversion of cover data that are stored as symbols of a specific cover scale to a decimal scale.
ORACON
A VAX-FORTRAN-77 program for conversion of decimal cover data to the so-called CORNELL condensed format needed for running application programs.

NAMCEK
This VAX-FORTRAN-77 program checks species names for correct spelling.

```
                 transect relevés          other relevés
                          |                    /     \
                          |                   /       \
                          |                  /         \
input:   TRANSASS.HBA(Husky)   EDITOR        VEGINPUT(VAX)
                          |          |              |
                   (*.HUS)        (*.DAT)        (*.FUL)
                          |          |              |
correction:        EDITOR(Atari)   EDITOR         EDITOR
                          |          |              |
                   (*.HUS)        (*.DAT)        (*.FUL)
                          |             \           /
conversion:        CONVERS(ATARI)         OPNACON
                          |                    |
                          |____________________|
                                     |
                            (*.VEG + *.OPN)
                                     |
storage:                         VEGINORA
                                     |
database:                         ORACLE
                                     |
selection:                        ORACLEX
                                     |
                            (VEGETATION.LIS)
                                     |
conversion:                        DECIM
                                     |
                                  (*.DCM
                                     |
                                   ORACON
                                     |
                                  (*.CND)
                      _______________|
                     |
Check:               |--NAMCEK-----(NAMCEK.OUT)
                     |
                     |        /     (ELLEN.OUT)
Ellenberg-           |--ELLEN------(*.EMR)
values:              |                     \
                     |                      \
Ordination:          |--DECORANA---(*.DMR)-COMBI---(*.CMB)
                     |          \__(*.DEC) /           |
                     |                    /       SILVISTAT
Clustering:          |--TWINSPAN---(*.TMR)             |
                     |          \__(*.TWI)\         figures
                     |                     \
Ordination:          |__CANOCO--(*.CAN)     \
                                    CLUTAB--(CLUTAB.UIT)
```

In capitals program names. Between brackets (*.DAT) resulting files with their extention are indicated.

ELLEN
A VAX-FORTRAN-77 program computing weighted mean values of Ellenberg indicator values for vegetation relevés.

COMBI
This VAX-FORTRAN-77 program combines output of ELLEN with machine readable output of TWINSPAN or DECORANA (*.EMR, *. TMR, *.DMR) in one file (*.CMB), that can be used for regression between Ellenberg values and axis of ordination. The file can also be used to plot figures with SILVISTAT.

Scheme for Processing of Tree Data

ARBODIGI
A GW-BASIC program on an OLIVETTI-PC for digitizing X- and Y- values of stem base and peripheral points of crowns in handdrawn plans and profiles on a CALCOMP digitizer tablet.

COREAREA.HBA
Husky BASIC program run on the hand-held computer for core area tree characteristics inventory, co-ordinate measurements and Dbh and height measurements with connected electronic calliper and altimeter.

PRECONTR
A VAX-FORTRAN-77 program that combines field form files with digitized plans to a so-called standard file (*.STF) that can be plotted with ARBOPLOT and checked by ARBOCONTR.

ARBOCONTR
A VAX-FORTRAN-77 program that checks a standard file (*.STF) on errors caused by unrealistic combinations of data from field forms and ground plan. The errors may have been caused by tree number interchange, wrong notes on the field forms, errors made in typing the field forms or digitizing the plans. The program outputs the file *.COR in which detected errors are sorted according to error classes. Together with a preliminary plot of the standard file with the program ARBOPLOT on the ATARI field station, computer errors can be checked in the field. If data come from direct HUSKI hand-held computer input a run of ARBOCONTR is redundant.

POSTCONTR
This VAX-FORTRAN-77 program unravels a corrected standard file (*.STF) into the field form files (*.FF1) and a file with plan co-ordinates (*.PLA) that can be loaded in the ORACLE database.

ORACLIN
This VAX/VMS DCL command procedure generates a SQL procedure which loads field form files (*.FF) and a file with plan co- ordinates (*.PLA) in the ORACLE database.

ORACLEX
This VAX/VMS DCL command procedure generates a SQL procedure which selects and sorts tree data from the ORACLE database on year, location of the core area on the position of trees in the core area and on other fields of a standard file such as tree number, species, Dbh, tree height etc. A standard file (*.STF) with selected tree data is the output of the program.

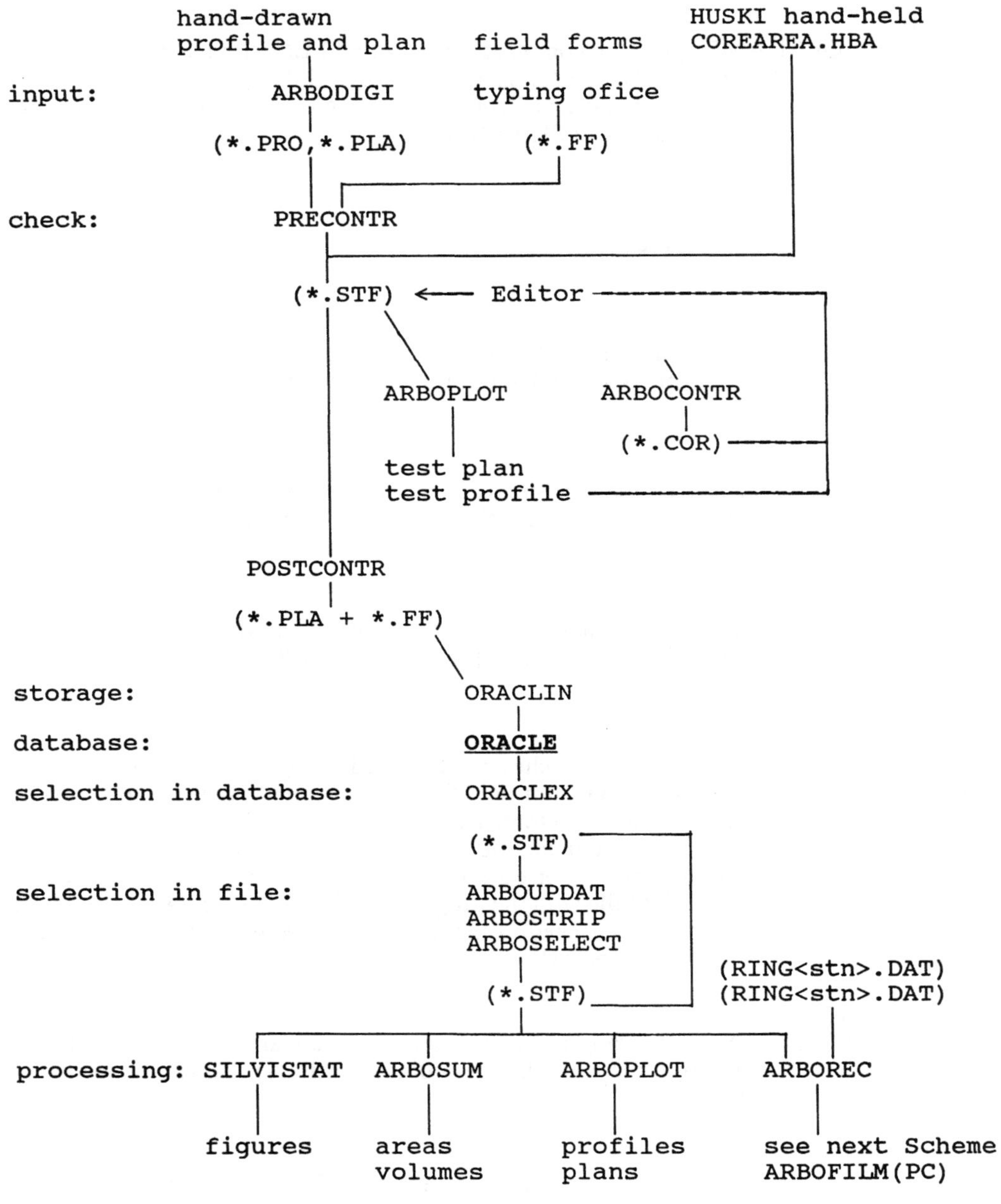

ARBOUPDAT
This VAX-FORTRAN-77 program selects the most recent data per tree number from a standard file (*.STF).

ARBOSTRIP
This VAX-FORTRAN-77 program selects all trees in a specified strip transect through the core area from a standard file (*.STF).

ARBOSELECT
This VAX-FORTRAN-77 program can provide selections from a standard file (*.STF) similar as can be made in ORACLEX. The advantage lies in shorter computation time.

SILVISTAT
A VAX-FORTRAN-77 program for plotting frequency diagrams, point plots, regression curves, making use of the program package GENSTAT.

ARBOSUM
VAX FORTRAN-77 program computing crown areas, real crown cover, crown volumes and basal areas per tree, per species, per 10 x 10 m block and per hectare.

ARBOPLOT.FOR
A VAX-FORTRAN program also GFA-BASIC version for ATARI and GW- BASIC version for OLIVETTI for plotting crown maps and profiles of forest structure. On PC computers also use of screen graphics.

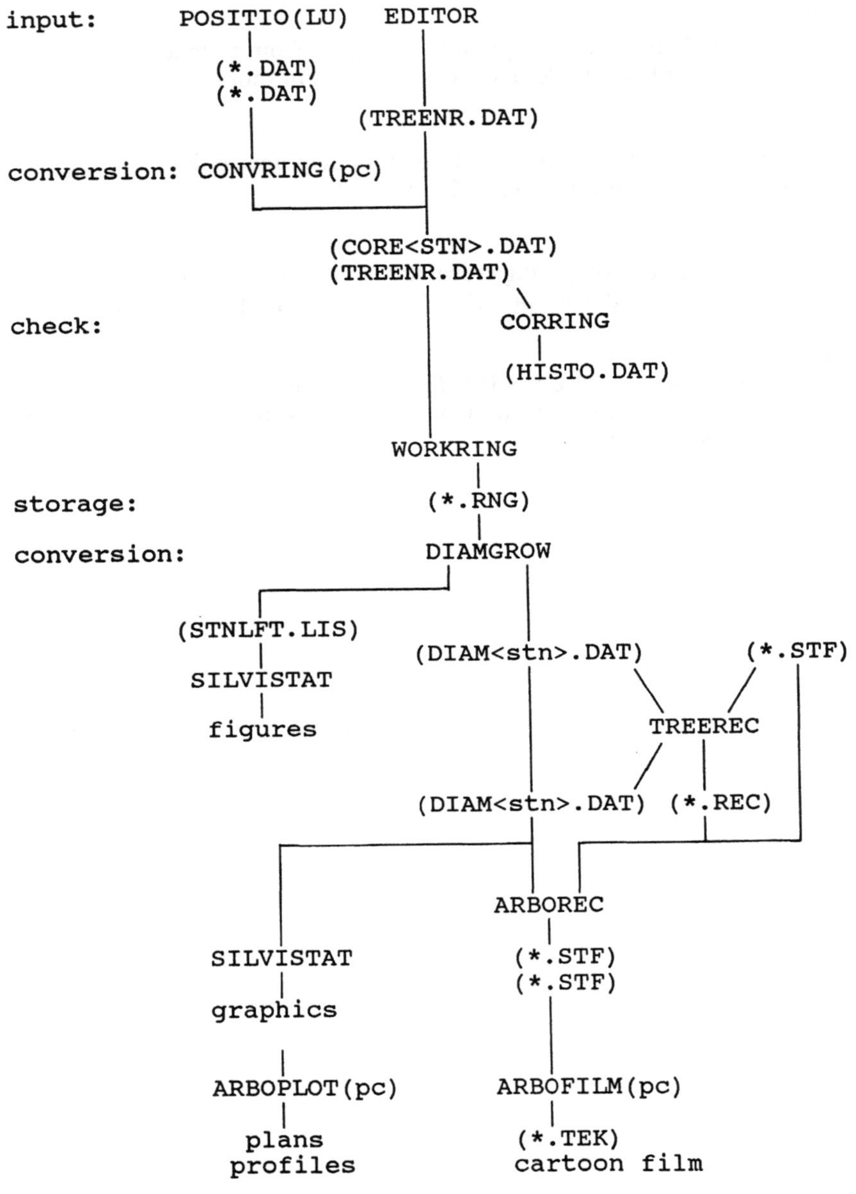

POSITIO
A FORTRAN program running on a PC in combination with a digitizer tablet for measuring tree ring cores (Agricultural University Wageningen, Department of forest ecology and silviculture). Tree rings widths per tree are written to *.DAT files.

CONVRING
A VAX-FORTRAN-77 program for conversion of *.DAT files of the program POSITIO to the format of CORE<STN>.DAT.

CORRING
A VAX-FORTRAN-77 program to check and convert raw tree ring data files. The program outputs histograms of sequential tree ring widths in files HISTO.DAT used to check and synchronize tree rings in different samples.

WORKRING
A VAX-FORTRAN-77 program for averaging coresponding tree ring widths of two samples of the same tree and calculating tree age. When necessary tree age is extrapolated. Tree rings are corrected for the real diameter of the tree and written to a file *.RNG.

DIAMGROW
A VAX-FORTRAN-77 program that computes cumulative diameter growth of sampled trees. Results are written to files DIAM<STN>.DAT.

TREEREC
A VAX-FORTRAN-77 that simulates a new standard file (*.REC) and tree ring file (DIAM<STN>.DAT) for trees that have died of which now tree ring sample is available and that have no actual crown dimensions. A new tree record in a standard file is generated according specified tree dimensions and new tree ring files are reconstructed from trees that have been indicated for reference.

ARBOREC
A VAX-FORTRAN-77 program that produces reconstructed standard files (*.STF) per year according regression of Dbh against all other tree dimensions using the present tree data (*.STF) and optionally reconstructed tree data (*.REC). The Dbh for a year is read from DIAM<STN>.DAT files.

ARBOFILM
With this GW-BASIC program the reconstructed sequential yearly standard files (*.STF) are used to draw plans and profiles on the computer screen which are saved as screen files (*.TEK). By sequential loading of the screen files a cartoon film can be presented. Optionally on a colour screen per species different colours can be specified.

Scheme of the Program and Procedure FOREYE

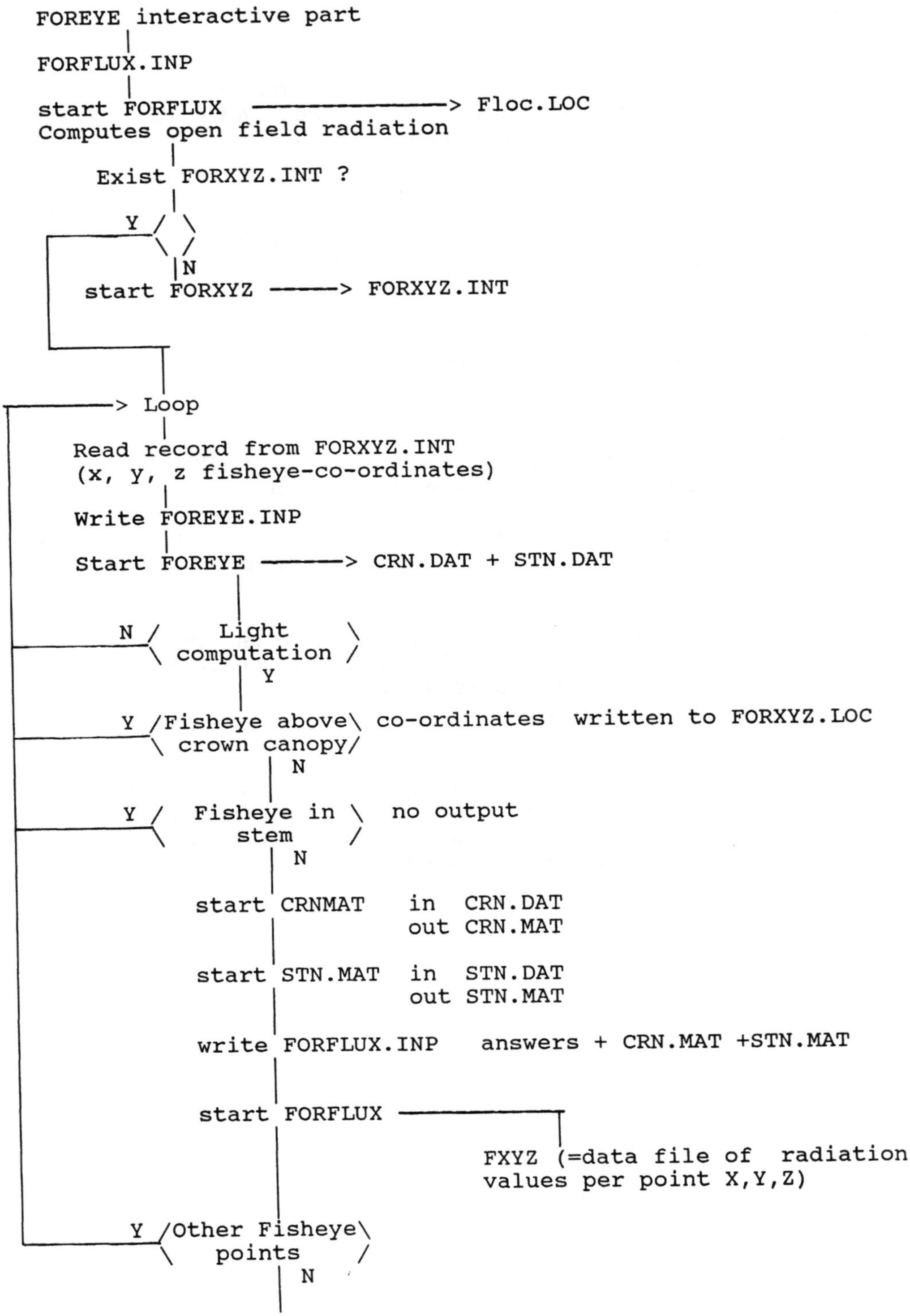

The program FOREYE is written as a VAX/VSM DCL (Digital Command Language) command procedure that subsequently starts VAX-FORTRAN programs. An interactive procedure FOREYE starts a non- interactive procedure SUBFOREYE in batch. The specifications given in the interactive part are written to a file FORFLUX.INP, which is used as input for other programs in the rest of the procedure. Where necessary during the rest of the procedure information from preceding to following programs is transferred via this file FORFLUX.INP as well.

FORXYZ
A program that computes a X, Y, Z, grid, which can be specified in the interactive part of the FOREYE procedure. At each point of the grid hemispherical light computations will be made. The program outputs a file FORXYZ.INT. If a file FORXYZ.INT, with specified points for fish-eye computations, already exists the program is not run.

FORFLUX
A program that computes radiation values from hemispherical coverage matrix files (CRN.MAT for degrees covered by crowns and STN.MAT for degrees covered by stems). The output is a set of data files per point in the X, Y, Z grid (specified in FORXYZ.INT), that contain daily radiation values. Each file FORXY.ZZ (for example at point (100, 30, 0 the file name is FOR10030_.Z0_) contains daily sun duration, diffuse light clear sky, diffuse light overcast sky, diffuse light total and total radiation. To prevent the program to run more than once for identical fish-eye points above the tree canopy without any hemispherical cover at all, at the beginning of the procedure, FORFLUX is run once for an uncovered hemisphere. The resulting values for radiation are substituted later on for all points without hemispherical cover that are registered during the run in the file FORXYZ.LOC.

FOREYE
A program that computes the vertical and horizontal degrees that delimit separate tree crowns and tree stems (see chapter VI). The program produces the files CRN.DAT for tree crowns and STN.DAT for tree stems (see Fig. 6.1).

FORPLOT
A program plots separate tree crowns and/or stems in a horizontogram from the files CRN.DAT and STN.DAT on a HP-plotter. There is an option to add sun tracks to the horizontogram (see Fig. 6.2).

CRN.MAT and STN.MAT
A program that converts single tree coverage data from CRN.DAT and STN.DAT to a hemispherical coverage matrix for all trees together.

READFLUX
A program that must be run after the SUBFOREYE procedure and combines a set of FORXY.ZZ files with radiation values of separate fisheye points that were specified in the grid file FORXYZ.INT to one file. This file per record contains X, Y, Z, item 1, item 2,.... The items can be the different radiation values computed in FOREYE averaged over a period of days to be specified. This file is converted to ARC/INFO format to produce maps of the different aspects of radiation.

```
CANOPIX interactive part
       |
CANOPIX.INP
       |
start CANOPIX ---------------+
       |                     |
      stop              binary file
                        CANOPIX.OUT
                             |
                        loaded in ERDAS
                             |
                        loaded in ARC/INFO
                             |
                           maps
```

The program CANOPIX is written as a combination of DCL (Digital Command Language) command procedure that subsequently starts VAX- FORTRAN programs. An interactive procedure CANOPIX starts a non- interactive procedure in batch. The specifications given in the interactive part are written to a file CANOPIX.INP, which is used as input for other programs in the rest of the procedure.

Forest Objects Under Study with the Monitoring System

A list of all forest objects included in the monitoring system in December 1988.

Core areas

Object	Owner
Kijfhoek	Duinwaterleiding Den Haag
Fontainebleau I en II	French State Forest Service
Bentheim	Fürst zu Bentheim (F.R.G.)
Ile de Rhina	Gemeente Rhinau (France)
Motketel	Kroondomein
Naardermeer oude kooi	NM
Norgerholt	NM
Otterskooi	NM
Weenderbos	NM
Bialowieza I en II	Polish state
Wilgenreservaat	RIJP
Keizersdijk	SBB
Schone Grub	SBB
Galgenberg (Amerongen)	SBB bosreservaat
Grootvenbos	SBB bosreservaat
Het Leesten	SBB bosreservaat
Het Quin	SBB bosreservaat
Lheebroekerzand (Dwingeloo)	SBB bosreservaat

Meerdijk (Spijkbos)	SBB bosreservaat
Nieuw Milligen	SBB bosreservaat
Pijpebrandje/Dikkenest	SBB bosreservaat
Schoorl	SBB bosreservaat
Star Numansbos (Gaasterland)	SBB bosreservaat
Tussen de Goren (Chaam)	SBB bosreservaat
Vechtelanden (Ommen)	SBB bosreservaat
Vijlnerbos	SBB bosreservaat
Zeesserveld (Ommen)	SBB bosreservaat
't Sang	SBB bosreservaat
Noordhout	Utrechts Landschap

Transects

Object	Owner
Marienwaard	Baron Verschuer
Anses wood (New Forest)	Forestry Commission
Beaulieu river (New Forest)	Forestry Commission
Berry wood (New Forest)	Forestry Commission
Bratley wood (New Forest)	Forestry Commission
Denny wood (New Forest)	Forestry Commission
Fritham plain (New Forest)	Forestry Commission
Hollands wood (New Forest)	Forestry Commission
Mark ash wood (New Forest)	Forestry Commission
Pinnick wood (New Forest)	Forestry Commission
Rushpole wood (New Forest)	Forestry Commission
White moor (New Forest)	Forestry Commission
Noorddijk	Gelders Landschap
Edese bos I en II	Gemeente Ede
Oevermansbos (Emmen)	Gemeente Emmen
Bekendelle	NM
Grevenmaat	NM
Herenduinen	NM
Leuvenumsebos	NM
Molenduinen I en II	NM
Wolfheze I en II	NM
Nijenbeek	Privat owner
Beerenplaat	SBB
Berkenvalei (2 x, Terschelling)	SBB
Broekhuizen I en II	SBB
Gerendal	SBB
Kroonpolder (Terschelling)	SBB
Schoonlo	SBB
Smilde (2x)	SBB
Ugchelen	SBB
Willinkweust	SBB
Galgenberg	SBB

Glossary

Architecture
: The visible, morphological, expression of the genetic blueprint of organic growth and development (Hallé et al. 1978).

Architectural code
: Code expressing the growth potential of trees, to what set or ensemble a tree belongs.

Architectural tree model
: Generalized growth program which determines successive phases of a tree (Hallé et al. 1978).

Canopy
: More or less closed upper layer of tree crowns.

Chablis
: A gap in the forest produced by the fall of a tree; the fall of the tree itself and the resulting forest damage (from the French) (Hallé et al. 1978).

Core area
: The area in the information system measuring 70 x 140 or 50 x 140 m where detailed information on position of trees and tree crowns is collected.

Coppice
: Woodland management system with trees cut at regular intervals

Crown area index (CAI)
: Total projected crown area above unit area of ground surface; a measure of canopy density (Hallé et al. 1978).

Crown projection
: Vertical projection of the crown periphery on a horizontal ground plane.

Eccentricity
: Eccentricity of the tree base, tree fork, tree crown base and top of the tree are expressed in x- and y-direction of the co-ordinate system as the ratio of distance from the point concerned to the centre of the crown and the crown radius.

Ecotope
: Combination of niche and habitat occupied by a plant (Hallé et al. 1978).

Eco-unit
: see Regeneration unit

Emergent
: An isolated tree standing above the average forest canopy (Hallé et al. 1978).

Ensemble
: see Set

Feature space
: Multidimensional space where vectors represent the values of coulor bands of a pixel from a digital remote sensing image.

Forest type area
: The area in which one forest type is studied every 10 years for those aspects concerning the hierarchical level of the silvatic unit.

Groundtruth
: Field data on objects seen on remote sensing images, necessary for interpretation of the remote sensing image.

Herbaceous transect area
: Area in which the herbaceous species, mosses and liverworths are described every 3 years.

Hierarchy
Here hierarchy in systems; a system is composed of subsystems, which in turn are composed of smaller subsystems.

Inversion point, morphological
The insertion of the lowest major limb of a tree above which all branch or reiterative complexes are progressively smaller (Hallé et al. 1978).

Inversion surface, morphological
Collective of inversion points of individual trees in a forest, indicative for that forest's status (Hallé et al. 1978).

Layer
The collective of plant life-forms that occupies a certain height interval.

Line transect
Transect drawing in which all trees are drawn of which the crown projection intersects or touches the midline of the transect.

Minimum area
Area needed to have a complete species composition

Minimum structure area
see Silvatic unit

Niche
The energetic position of a species in a community (Hallé et al. 1978).

Periphery
Greatest width of the crown, in fact the crown projection.

Peripheral point
In the tree model the four extreme points in positive and negative x- or y-direction of a coordinate system are called peripheral points.

Peripheral plane
Plane in the tree model at the peripheral height level in which the peripheral points stretch up quarter ellipses.

Peripheral height
The averaged height of the peripheral points of a tree.

Permanent plot
Plot that is permanently marked in the field and is reassessed in time.

Phase
State in a cyclic successional series that is characterized by the age of the stage of which it is part.

PNV
see Potential Natural Vegetation

Potential Natural Vegetation
The species composition that would develop, under present existing abiotic circumstances, without human interference and within a time span over which the vegetation can reach its end state (Tüxen 1956). A stand without progressive succession is already its own PNV.

Potential tree
Tree with potential for future expansion within the forest mosaic (Hallé et al. 1978).

Profile drawing
Drawing of side view of a line or block transect.

Reiteration
Development of shoots outside the normal expression of the architectural model of a tree, as a specialized environmental response (Hallé et al. 1978).

Regeneration unit

The regeneration unit or eco-unit is: "every surface on which at one moment in time a vegetation development has begun, of which the architecture, ecophysiological functioning and species composition are ordained by one set of trees until the end" (Oldeman 1983, 1989).

Relevé

Inventory of plant species and their cover of a plot.

Sapling

Young tree (here less than 5 cm Dbh)

Seedling

Tree less than 1 year old

Set (or ensemble)

Group of trees within a forest with equivalent developmental potential (e.g. potential set, set of the present and past) (Hallé et al. 1978).

Site mosaic

Complete mosaic consisting of different sites, with different abiotic conditions, characteristic for a region.

Stand

see Regeneration unit

Stage (of cyclic succession)

State in a successional series determined by its own characteristic species composition.

Strip transect

Rectangular transect plot in which only trees with their stem base within the plot are included.

Structural ensemble

Subdivision of the set of the present according to height (Hallé et al. 1978).

Sylvatic mosaic or unit

The minimum surface at any one site needed for all architectural growth phases of all locally possible eco-units (Oldeman 1989).

Synusiums

Groups of plant populations that belong to one life form or are separated from other populations in time or space or by a fundamentally deviant use of the site (Barkman 1973).

Thinning

Cutting of trees to favour other trees in their growth Tree of the present Dominant tree which codetermines forest architecture, but without further capability for expansion (Hallé et al. 1978).

Tree of the past

Over-mature, decaying or damaged tree in the process of elimination from the forest mosaic (Hallé et al. 1978).

Uprooting

The fall of a tree, where the root system is uplifted from the soil and the resulting mound-and-pit micro-relief.

Woody transect area

Rectangular 10-m-wide area in the centre of the core area, in which a side view is drawn.

Zenith

Top of the hemisphere vertically above the observers point.

a, b	Half axix of an ellipse in the crown projection
B	Stem base
Bax	X co-ordinate of stem base
Bay	Y co-ordinate of stem base
Baz	Z height co-ordinate of the stem base
C	Base of the crown
Crx	Crown radius in X direction
Dbh	Diameter breast height of the stem
E	Exponent of the vertical crown curves
eBx	Eccentricity of the tree base in X direction
eTy	Eccentricity of the tree top in Y direction
F	First branching fork of the stem
HC	Height of C
HF	Height of F
HP	Height of P
HT	Height of T
Mx	Centre point of the crown in transverse section in X direction
MPn	Peripheral point on the mound of an uprooting
Pn	Peripheral point at the height of the crown with the greatest width
P1x	X co-ordinate of the first peripheral point
P1y	Y co-ordinate of the first peripheral point
T	Top of the crown
Tx	X co-ordinate of the top of the crown
Ty	Y co-ordinate of the top of the crown
V	View point of the fish-eye photograph

Species Index

Subject Index

FSC
www.fsc.org
MIX
Papier aus verantwortungsvollen Quellen
Paper from responsible sources
FSC® C105338

If you have any concerns about our products,
you can contact us on
ProductSafety@springernature.com

In case Publisher is established outside the EU,
the EU authorized representative is:
Springer Nature Customer Service Center GmbH
Europaplatz 3, 69115 Heidelberg, Germany

Printed by Libri Plureos GmbH
in Hamburg, Germany